Errata

Because of an indexing error, members of the author's family are incorrectly listed in the index under the name Nielsen. With the exception of Margrethe Nielsen, their names should read Schmidt-Nielsen.

The
Camel's
Nose

Knut Schmidt-Nielsen

The
Camel's
Nose

Memoirs of a Curious Scientist

Illustrations by Kathryn K. Davis

ISLAND PRESS / Shearwater Books
Washington, D.C. · Covelo, California

A Shearwater Book
published by Island Press

Copyright © 1998 Knut Schmidt-Nielsen

All rights reserved under International and Pan-American Copyright Conventions. No part of this book may be reproduced in any form or by any means without permission in writing from the publisher: Island Press, Suite 300, 1718 Connecticut Avenue NW, Washington, D.C. 20009.

Shearwater is a trademark of the Center for Resource Economics.

LIBRARY OF CONGRESS CATALOGING-IN-PUBLICATION DATA
Schmidt-Nielsen, Knut, 1915–
    The camel's nose: memoirs of a curious scientist /
    Knut Schmidt-Nielsen
    p.   cm.
    Includes bibliographical references and index.
    ISBN 1–55963–512–6 (alk. paper)
    1. Schmidt-Nielsen, Knut, 1925–   2. Physiologists—biography.
    I. Title
QP26.S33A3   1998
571.1'092—dc21   98–5969
[B]   CIP

Printed on recycled, acid-free paper

Manufactured in the United States of America

10 9 8 7 6 5 4 3 2

Contents

Part III
Changes [1962–1975]

Part IV
Measuring a Life [1975–1995]

Preface

THIS IS A personal story of
a life spent in science. It tells about curiosity, about finding out and
finding answers. The questions I have tried to answer have been very
straightforward, perhaps even simple: Do marine birds drink sea
water? How do camels in hot deserts manage for days without drinking
when humans could not survive without water for more than a day?
How can kangaroo rats live in the desert without any water to drink?
How can snails find water and food in the most barren deserts? Can
crab-eating frogs really survive in sea water?

These are important questions. The answers not only tell us how
animals overcome seemingly insurmountable obstacles in hostile environments; they also give us insight into general principles of life and
survival.

I have also examined and questioned my own life—both emotional
and intellectual—in environments that have not always been friendly.
Over the better part of a century, I have lost colleagues and family
members, witnessed breathtaking discoveries and changes, and found
great satisfaction in and beyond science.

I have often wondered what made me a physiologist and not an engineer or a carpenter or a physician. I could probably have managed reasonably well in any of those fields. But I was always curious about animals, and because my father eventually permitted me to choose my own way, I have enjoyed the excitement of a life spent in finding out how animals work.

The
Camel's
Nose

Part I

How It All Came About

[1915–1946]

Of Salt and Science

IT HAS BEEN said that the primary function of schools is to impart enough facts to make children stop asking questions. Some, with whom the schools do not succeed, become scientists. I never made good grades in school. At times I nearly failed, and I never stopped asking questions. Indeed, I have spent most of my life asking questions and finding answers to how animals manage in the world around us.

Perhaps incurable curiosity is in the genes. My grandfather, an engineer who built railroads in Norway, was curious about the flounders that occasionally showed up in fresh water upriver from my hometown of Trondheim, a port city in central Norway. He wanted to know if this saltwater fish could survive for any length of time in fresh water, so in 1903 he released over one thousand flounder hatchlings in a large inland lake.

The fish survived, and for years afterward, startled fishermen occasionally snagged a grown flounder in the lake. However, the fish were unable to reproduce in fresh water and never became permanently established there.

As far as I know, my grandfather's test was the first experiment conducted in what we now call osmoregulation, the processes by which animals regulate the amounts of water and salts in their bodies.

Some years later, my father, a chemistry professor, examined a large number of saltwater fish and found that the salt concentration in their body fluids was only about one-third of the concentration of salt found in the surrounding sea water. He also measured the concentration of salt in whale urine to find out how these mammals manage to secrete the salts ingested with their food—a necessity for survival—but the results were inconclusive.

During my own work in physiology I have been curious about similar problems—how animals meet the challenges of their environment, how they adapt to life in the sea or in deserts. At first glance, these environments may seem very different, but from a physiological viewpoint, fresh water is no more freely available in the sea than in the desert.

How I came to be a scientist is a story that necessarily begins with my parents and my childhood.

My Swedish mother was born Signe Sturzen-Becker in 1878. One remote ancestor—a Robin Hood figure named Klaus Störtebecker—became a legendary folk hero in northern Germany, operating as a pirate against the Hanseatic League and its economic domination of the North Sea countries, where he distributed his loot to the poor and powerless. He was betrayed and beheaded in front of the Hamburg City Hall in 1401. Mother's immediate forebears were writers and scholars. Her grandfather studied early Greek poetry, and her father published a comprehensive grammatical analysis of early English dialects.

Mother's intelligence and energy were manifest early. The girls' school she attended considered mathematics a subject unsuitable for women. At sixteen she decided she wanted a more demanding education, and on her own she read the two years of math courses required for matriculation in physics at the Institute of Technology in Stockholm. She studied under Svante Arrhenius, the Swedish chemist who in 1903 was awarded the Nobel Prize for his theory of electrolytic dissociation by showing why salt solutions conduct electricity. She was excellent at mathematics, and Arrhenius occasionally sought her help with particularly difficult problems of integration. She was the third woman in Sweden to obtain a doctoral degree and the first to do so in physics.

My memory is of a slim, energetic woman whose dark, curly hair had turned prematurely gray and then white. She had a knack for explaining things to a curious boy. While she worked or supervised the maids in the kitchen, she told me about the physics of food and cooking—heat, boiling, eggs and coagulation, how olive oil forms an emulsion with egg yolk that we call mayonnaise, how cream is an emulsion of fat droplets in water, and conversely, how butter is an emulsion of water in fat. Her explanations were factual and never glib, and I absorbed much information and common sense from her.

My Norwegian father, Sigval Schmidt-Nielsen, was born in 1877 and grew up in Trondheim. His father, my grandfather, had been an engineer engaged in the construction of the first railroad from the capital, Christiania (now Oslo), across the mountains to Trondheim. With the completion of the railroad, Grandfather ceased designing bridges and became inspector of fisheries in Trondheim, where he also served as director of a small fisheries museum. His interest in biology led to his experiments with transplanting flounders to fresh water.

My grandfather had been born Olaf Ludvig Nielsen. Norwegian names ending in *sen,* meaning "son of," are very common because for centuries men were identified simply by adding *sen* to their father's first name. As urban populations increased, more and more people ended up with identical family names. Eventually some therefore chose to distinguish themselves by combining two names with a hyphen. My grandfather added his mother's family name, Schmidt, to that of his father, Nielsen, to produce a more distinctive family name.

I remember my grandfather as a white-haired old gentleman who every day took a walk around town, chatting with good friends and seeing to it that important matters had been properly attended to.

As a child I perceived my father as a strict man who ruled his house and family with absolute authority. Most would agree that he was a commanding presence, heavyset and of more than average height, with gray hair and mustache and jet-black eyebrows. As he grew older, his hair turned white, but his striking eyebrows remained as black as in his youth. My perception of him as stern and distant kept me from learning about his life and the ambitions that drove him, a young chemical engineer from a minor institution in Norway, to seek an advanced education in Germany, Belgium, and Switzerland. Only as I have aged have I realized that he would have liked to share with me more about himself and his life.

Father had been awarded a doctoral degree at the University of Basel in 1901, when enzyme research represented the cutting edge of scientific activity. The brothers Eduard and Hans Büchner had made the sensational discovery that cell-free yeast extracts still contained all the enzymes necessary for alcoholic fermentation, discoveries for which they received the Nobel Prize in 1907. For his degree Father analyzed the role of enzymes in the maturing process of salted herring. His lifelong interest in food was reflected in his later work on food technology, nutrition, and vitamins and, after his retirement, a dictionary of food and gastronomic terms.

Father's interests spanned many other fields. While working at the Institute of Physiology in Liège, Belgium, he published a paper refuting the current belief that the two atria of the heart contract simultaneously. With the aid of dog hearts and a mechanical device called a kymograph, in which a pointer writes a fine trace on a revolving drum covered with sooted paper, he showed that the right atrium contracts a few thousandths of a second before the left one.

In 1904 my father was appointed a fellow in biological chemistry at Stockholms Högskola (now the University of Stockholm) and a lecturer in medical chemistry at the Karolinska Institute. It was here that he met Arrhenius' student and collaborator, Signe Sturzen-Becker.

After their marriage in 1907 my parents moved to Oslo, where Father was appointed lecturer in physiology at the university. In 1908 he became chief chemist for the Directorate of Medicine, and during the next five years he carried out a wide variety of studies of food purity, adulteration, use of illegal preservatives, lead and arsenic poisoning, chemical pollution, and the chemistry and bacteriology of drinking water. In addition, he worked as a consultant to the Patent Office.

Much later Father became a consultant to the Parliament, helping to revise and write legislation pertaining to food processing and distribution. He was especially interested in adulteration and chemical treatment of food and sought to ban an array of preservatives and undesirable additives used in the canning industry, as well as chemical treatments of flour to improve its baking qualities. He maintained that food colorings were merely cosmetic, in no way improved quality or taste, and might be harmful and even carcinogenic.

His distinguished appearance, immense knowledge, and convincing arguments resulted in the passage of legislation quite advanced for that time. But to his great disappointment, he failed to get a ban imposed on artificial dyes in processed sausages. The meat industry insisted that because housewives expected sausages to be red, they would not buy uncolored sausages. Father argued that if all sausages were produced without artificial colorings, people would probably eat the same quantities of sausages. However, in the end the meat industry's arguments prevailed.

In 1913 Father was appointed professor of technical organic chemistry at the newly established Norwegian Institute of Technology in Trondheim, his hometown. I was born on September 24, 1915, the youngest of four children. My sister, Astrid, was two, and my brothers, Tore Patrick and Klas, were four and six, respectively.

Our house, on an unpaved road at the outskirts of town, was set on a hilly, rocky lot with birch trees cloaked in shiny, paperlike bark and mountain ash that in the autumn held large clusters of bright red berries. From the second floor we could look out over the town and the fjord, where the wind often whipped the dark gray water into foaming waves. On the other side of the house were open fields and wooded areas.

Father had designed our house, and the family moved in the year I was born. Upstairs were my parents' bedroom, two rooms for live-in maids, a large bedroom for Klas and Patrick, and a smaller room for Astrid. Clearly, the house had not been designed with a fourth child in mind, and Astrid and I shared her room until we were of school age. Later, when we had only one live-in maid, Klas got a room of his own, and I shared a room with Patrick.

In addition, Grandfather's sister, Aunt Margrethe, came to live with us every winter, sleeping in a separate room in the attic. She was white-haired, old-fashioned, neat, and always dressed in black. Father helped

Aunt Margrethe and other relatives financially, but this fact was never mentioned; helping poor relatives was an unspoken obligation.

With parents, an aunt, and two maids in the house, there was an ample supply of adults to supervise my upbringing. One maid tried to teach me to use a handkerchief. "Sniffling," she said, "will cause worms in your nose, and they are terribly painful to remove." I doubted her dire prediction, and sniffling remained a convenient solution for a runny nose when no adults were around.

Once a month a washerwoman came to do the laundry. She boiled our sheets and other linens in the basement in a huge iron kettle heated with firewood, rinsed them in large wooden tubs, and then hung them to dry on lines strung between trees or, when it rained, on clotheslines in the attic. In winter the wet sheets that hung outdoors froze to stiff plates, like boards of plywood. If they were left hanging, the ice would slowly evaporate until the sheets grew soft and dry, ready to be taken in and folded. Thus I learned early that some solids could evaporate; much later I learned the proper scientific term for the process—sublimation.

A dressmaker came to the house to make clothes for Astrid and Mother. Ready-made suits for men were unheard of at that time. Instead, a man went to a tailor, who took measurements and discussed material and price. The suit would be used for years, and when it was worn out it would be taken to a tailor, who opened all the seams and put the pieces together again, turned inside out. The result was a suit that looked almost new, except that the position of buttons and buttonholes was reversed. My first suit, acquired when I was thirteen or fourteen, had been remade in this way from one of Grandfather's. I was terribly ashamed to wear it, certain that everybody would notice that the buttons were on the wrong side.

Father came home for dinner at three in the afternoon. After a nap he went back to work and didn't return until late in the evening. Most weekdays we had fish—cod, mackerel, deep-water rosefish, and many others, depending on the season. Usually we had meat on Sundays and once or twice during the week. Father was very particular about the freshness of the fish and often commented that fish kept on ice for a day wasn't properly fresh. Sometimes I saw fish in the kitchen so fresh it was still flopping around on the counter.

Our repertoire of vegetables was limited. The winter's supply of potatoes arrived in the fall and was stored in large bins in the cellar to last until the following summer. The common vegetables included cabbage, carrots, and sometimes kohlrabi or turnips. Only in late summer did we occasionally have lettuce, tomatoes, or cucumbers.

As a child I simply took for granted the interesting conversations we often had at the dinner table. When there was something to look up, an unusual word or a plant we wanted to know more about, Father would send Astrid or me to fetch a book or a volume of the encyclopedia from his library. In this way we became used to looking up answers when there was something we didn't know. This habit has served me well ever since, and my own children grew up accustomed to reaching for a dictionary whenever there was something they didn't know.

Father's knowledge of nutrition and vitamins had one inescapable consequence: during the dark winter months a spoonful of cod liver oil was administered daily to each child. The taste was dreadful. "Take a deep breath and you won't notice," the adults advised. It didn't work, but I discovered a trick that I later taught my own children. Before the oil comes into your mouth, exhale as far as possible. Then, after swallowing the oil but before inhaling, immediately swallow some food or drink, and the taste is hardly noticeable.

On mornings before school, we children always had oatmeal porridge, often badly cooked and sometimes scorched if the maids were late and had turned up the gas heat to hurry the pot. Any leftover porridge was cut into thick slices, which we had to eat the next morning, fried in margarine.

Father ate his breakfast after we went off to school. On Sundays, when we all had breakfast together, we never had oatmeal. In the light of later discoveries it appears that our dislike for the stuff was justified. Oatmeal porridge is little more than water thickened with starch, and even worse, oatmeal contains phytic acid, which binds the calcium in food. Milk should provide ample amounts of calcium for growing children, but adding oatmeal to the milk impedes calcium absorption. Unfortunately, all this was unknown when I was growing up.

Our milk was delivered by a farmer who came every morning with his horse-drawn cart and scooped milk from a huge pail into our smaller kitchen pail. We poured the milk into large bowls and left them undisturbed until the cream could be skimmed off and used for coffee and desserts. I knew no other kind of milk until pasteurization and bottled milk were introduced, when I was about ten. Pasteurized milk tasted odd to me, but it was not as bad as the milk we had a few years earlier during an epidemic of scarlet fever, when Father, who knew that raw milk could be a factor in the spread of epidemics, ordered that all our milk be boiled.

The weeks before Christmas brought a bustle of kitchen activity; baking went on for weeks, and tins full of cookies were saved for the holidays. The local butcher delivered half a dressed hog, and then the kitchen was hectic with preparation of roasts, side ribs, pickled pig's feet, liver pudding, and much else. With assistance from Aunt Margrethe and the maids, Mother made lengths of salamilike sausage, preferably from horse meat, which many said was the best. The sausages were sent out for smoking, and then hung in the attic to dry further.

There was nothing especially unusual about eating horse meat, which was often found in the butcher shops. One Saturday afternoon when the maids were off, I was in the kitchen with Mother, listening and asking questions as she prepared a large piece of meat for Sunday dinner. The kitchen was cool; outside, the late autumn light fell across our neighbor's large wooded and rocky yard. Mother put the meat in a pottery crock, adding laurel leaves, pepper, herbs, and a dilute mixture of vinegar and water to make a dark, murky marinade. As she put the crock away in the larder, she said, "By the way, when Grandfather comes for dinner tomorrow, don't mention that we are eating horse meat."

"Why?" I asked.

"Some people, like Grandfather, will not eat horse meat," she explained. "But it is hunting season, and he will assume he is eating venison."

Then I had a clever idea. "When Grandfather eats the meat tomorrow, he will enjoy it and praise it," I said, "and then we will tell him what it is, and he will understand how dumb it is not to eat horse meat."

"Oh, no," Mother said, turning toward me. "If you make somebody feel like a fool, he gets very angry."

This was my first lesson in applied psychology.

Like my father, I enjoyed most food and still do today. In my preteen years there was no limit to my capacity. One Christmas we had an old-fashioned plum pudding for dessert, sweet and moist and lovely to look at, made with suet, currants, and raisins. The pudding, wrapped in a linen towel, was boiled for hours and brought to the table steaming hot. Father poured cognac over the hot pudding and lit it; the flames leapt up and left a lovely flavor behind. I had already eaten my fill of the main course, but I had several helpings of the pudding. Then I downed a huge glass of cold water and leaned back in my chair, satisfied.

My brother Patrick looked at me and explained, "At body temperature suet remains melted, but once it is congealed by cold water, it doesn't melt again and remains in your stomach as a big undigested lump." That night I tossed around in my sleep, imagining the lump of solidified suet like a rock in my stomach.

Father liked to entertain and would often invite friends, visiting scholars, or colleagues to a dinner party. Sometimes my sister, Astrid, and I watched the guests arrive, looking out the window in her upstairs room, kneeling on the floor with our noses just above the windowsill. Norwegians knew that for a formal dinner they could politely arrive as much as five minutes late, but ten minutes would be very impolite, for it would upset the hostess and ruin the well-planned kitchen schedule. A Swedish guest would be more punctual and probably arrive a couple of minutes early, pull out his pocket watch, walk up the street for a short distance, turn around, look again, and perhaps go for another short walk. Exactly on the hour his finger would be on the doorbell.

The guests would gather in the living room for no more than a few moments of conversation; then Father would open the wide doors to the dining room, and the guests would stream in. Each place setting had a name card, china, and silver, and four or five wine glasses. Food and wine came in the traditional order—soup, fish, meat, sometimes cheese, and finally dessert.

My brothers and I often served at these dinners. Father taught us how to balance a serving platter, change plates from the right, serve food from the left, and pour the wine from the right, giving the bottle a slight twist to avoid spilling. He seemed friendlier on these occasions, teaching us the proper techniques without issuing strict orders. In retrospect, I believe that he was proud of his sons and cherished showing us off to his friends.

School

WHEN ASTRID and I were small, we often played together. Although she was only two years older, she seemed full of wisdom and knowledge. She taught me the names of flowers and trees and which berries I could safely eat and which were poisonous. When she entered school in 1920 I felt abandoned, and when she went to school even on my birthday, I considered it a grave injustice.

Astrid didn't really teach me to read, but I rapidly picked it up anyway. One rainy day we were lying on the floor facing each other on opposite sides of a huge sheet of paper. I copied her letters, turning them around in my mind and writing them right side up. However, I flipped the letter *L* incorrectly so that it looked like a *T* with the left part of the top bar snipped off.

At that point Mother came in, looked at our work, and commented, "Knut, your *L* isn't quite right." I was sure I had faithfully copied Astrid's writing, and after looking at it again, I knew my letter was like hers. I was torn between the authority of these two females. Finally, Mother's prevailed, and I decided to put the horizontal line at the bottom of the vertical line. It has remained there ever since.

A year later I entered first grade with about twenty-five other boys; girls were in another part of the school building. We learned to write cursive script, but even though I was originally left-handed, I had no difficulty using my right hand. I have continued writing right-handed, but when I lecture to a class, I may pick up the chalk and write with my left hand if that side of the blackboard is more convenient. I can carve a piece of meat with either hand; I wield a hammer or an ax better with my right hand; but cutting with scissors or careful whittling of a piece of wood requires the left. I feel I can safely maintain that I am ambidextrous.

When I was in first grade, the other boys in the class were learning to read. The teacher told them first to say each letter in a word out loud and then to put them together to make the word. The next word was done the same way—letter by letter, and then the word. When the teacher called on me to read, I naively started reading quickly the way I was used to, without spelling out the letters. She quickly interrupted me, shouting, "Not so fast, not so fast!" I felt humiliated that I hadn't understood the proper way to read in first grade.

In second grade I was sick when school started. One day after my return the teacher called on me to answer a question. I said that I didn't know the answer because I had been absent. "But haven't you read your home assignment?" she asked. Not having been there when she first explained it, I hadn't heard about homework; instead I had kept up for weeks by listening carefully in class.

Events like this became a recurrent theme—whenever I had missed an explanation, I felt like a failure. Because I was the youngest child, often no one thought to tell me what my brothers and sister already knew and took for granted. In raising my own children, I have tried to remember that matters that seem trivial to grown people can be very important to a small child.

TRONDHEIM IS located on a wide fjord at the mouth of the river Nidelven. In 996 it was established as a *kaupang,* an old Nordic word for a town of commerce, which in modern terms means that it received its charter as an incorporated city.

When Christianity was introduced into Norway about a thousand years ago, the new faith encountered fervent resistance, with fierce battles between Christians and heathens in many parts of the country. In 1030 King Olav, a Christian leader, was killed in combat. Loyal friends escaped to Trondheim with the king's body, which they concealed in the bank of the river. Months later, when the body was uncovered, it had not decomposed, and the hair, beard, and nails had continued to grow. As a result of these miraculous signs, the king was sanctified as Saint Olav. Nearby a spring started flowing, and a chapel was built around it.

The town rapidly became a destination for Nordic pilgrims. In 1161 construction began on a gothic cathedral to replace the chapel. It was intended to match the most beautiful cathedrals of England and France, but in the late Middle Ages the edifice suffered from several fires and eventually fell into ruin. The walls collapsed, and stones were used for construction elsewhere. At one time Swedish troops occupying Trondheim stabled their horses in the building. In 1869 restoration was begun, and reconstruction of major sections was complete by 1930, the 900th anniversary of King Olav's death.

My school, called the Cathedral School, was founded by the church six hundred years ago but is now a public institution. Because of the nominal affiliation with the church, all students and teachers were required to attend special monthly services in the cathedral. Although I had no interest in religion, it was pleasant to be free from school for an hour and to listen to the magnificent organ, which was often played by prominent organists from elsewhere in Europe. Organ music still moves me deeply.

Trondheim lies 500 kilometers north of Oslo, which is at the same latitude as the southern tip of Greenland. In winter that far north, darkness prevails. The sun barely rises above the horizon for about an hour at midday; on cloudy days, electric lights are needed even at noon. We walked to school in darkness at eight in the morning and returned home in darkness about two. But that never bothered us; we were used to winters and skiing in darkness, and with a clear sky and stars it was easy to find the way home. After a good snowfall we could even ski to school until the traffic of horse-drawn sledges ruined the snow.

Spring would arrive in April with melting snow and water everywhere. By late May it might be warm enough for swimming, although the water was only about 12°C (54°F). In summer the water might reach 18°C (64°F), which we considered warm.

I was only moderately good on skis and a mediocre swimmer. My brothers were much better at sports, and I felt I could never match their skills. Faced with difficulties, I would give up. When I was five or six, for example, I decided to try ice skating. In those days skates were attached to the boots with leather straps. With a pair of skates in my hand, I walked alone through the darkness to a small frozen pond. My fingers were stiff with cold, and I couldn't fasten the skates on tight; they were so wobbly I couldn't stand, let alone move on the ice. I stubbornly refused to try again, and later, when my school class went skating, I simply said I had no skates. I was twenty-five and living in Denmark before I tried ice skating again; then I learned quickly and found it easy.

As a child I was terrified of dogs and horses. When the neighbor's black-and-white setter jumped up and licked my face, I panicked. Horses grazed in a nearby fenced area, but I barely dared cross a corner of the field for fear they would come after me. There were also reportedly bums whom Astrid and her friends perceived as a danger, although we rarely saw any. Nevertheless, when it was whispered that someone had seen a drunk, we made extensive detours. However, I don't recall being afraid of the dark or of ghosts, although we heard many scary ghost stories. Reality seemed more dangerous. Given my sheltered childhood, it is difficult to understand why I was so fearful.

One morning when I was eight I awoke with pain and a tender abdomen. Father immediately suspected appendicitis and called the surgeon who lived up the street. He stopped on his way to the hospital and confirmed the diagnosis but thought that we could wait and see how it developed. However, Father wanted me operated on immediately. The year before, Patrick had nearly died from peritonitis, and Father didn't want to risk another perforated appendix. An ambulance took me to the hospital, and I was in surgery the same morning.

During my eight days in the hospital the German biochemist Gustav von Embden walked a good twenty minutes to visit me every day. Embden was working in Father's laboratory that fall and was one of the kindest, gentlest persons I encountered as a child. He spoke no Norwegian, and I understood only a few words of German, but when he came to sit by my hospital bed for a few minutes I felt comfortable and secure. Years later, Astrid spent a year with the Embden family to learn German, and she reported similar warm feelings for this famous scientist whose biochemical research was crucial to our understanding of phosphate metabolism in muscles and what has come to be known as the Krebs cycle. During a bicycle tour with a friend through Germany when I was seventeen, I stopped in Frankfurt to see Embden, only to learn that he had died a few days before.

Mother also came to the hospital every day. She brought me a book titled *Indian Folktales,* which was a collection intended for adults and had no illustrations. I read the lengthy stories in a couple of days, but they bored me. What I really wanted, I said to Mother, was *Wild Animals and Their Ways,* by Samuel Baker, which was listed on the back of the book I was reading. Mother didn't think the book was suitable for me, but I insisted, and she brought it the next day, my eighth birthday.

The book opened a new world for me. I was totally absorbed by the tales of lions and elephants in Africa, of hunting tigers in India and wild boars in Turkey, of the world's fastest animal, the cheetah, and of hippopotami that left their murky rivers at night and gobbled up watermelons in the natives' gardens. Years later I learned that Sir Samuel Baker was famous for his exploration of the Upper Nile and the discovery of Lake Albert.

Later I spent hours looking through books in Father's library. One of my favorites was the huge encyclopedia; I could open any volume at random and read about something new. One day I found the recipe for gunpowder: 75 percent sodium nitrate, 15 percent sulfur, and 10 percent powdered charcoal. Soon I could produce a good bang by putting a little heap of powder on a rock, placing a smaller rock on top, and dropping a heavy rock on the first two.

When a new house was built across the street from us, the basement was blasted out of solid rock, which gave me an excellent opportunity for a big bang. After the workmen left for the day, I would put powder at the bottom of the excavation and drop a big rock from several meters above. This technique was safer than when I stood with my feet next to the explosion, and I could increase the charge.

One Sunday when I had produced some magnificent explosions, a neighbor was passing by and exclaimed, "Are they really blasting on a Sunday?" I shrugged and innocently said I hadn't heard anything.

In another experiment, I constructed a small brass cannon with a bore of about a centimeter. I made a fuse from filter paper, drenched with sodium nitrate and then dried. The fuse didn't seem to burn well, so I tried again with another match. I didn't get my hand out of the way, and when the cannon finally went off, a sharp flame flashed from the fuse hole, taking the skin off my left thumb. When I came home with the bare flesh showing, I told Mother that I'd had an accident at the neighbor's stone wall, but I refused to say exactly how. Later, when Mother heard from other children that I had fired a cannon, it was too late for her to get as angry as I thought I deserved. The burn healed slowly but left no permanent mark.

My enthusiasm for experimentation grew as I made friends with a new boy, Per Stensrud, who entered school in the middle of third grade. Per's father had bought the Britannia Hotel, the largest and most elegant in Trondheim. After school Per and I often walked to the hotel, where we were allowed the run of the place. Operating a large hotel with three restaurants required impressive technical services—kitchens, laundry, repair shops, and boiler rooms to provide steam and hot water. I found it fascinating and especially loved the kitchen, where the staff sometimes spoiled us with big portions of ice cream.

One year Per suggested we breed rabbits. We were eager to find out how they mated, but since he had only one rabbit, we didn't learn much. Nevertheless, our friendship grew. I built a telegraph system so we could communicate between our houses. To make a buzzer I needed an electromagnet and a circuit breaker. I wound insulated copper wire around an iron core for the electromagnet, improvised a telegraph key by cutting strips from a tin can and connecting the parts, and used flashlight batteries for power.

The system worked well when we sat on the floor in my room and sent messages to each other in

Morse code. But when we actually connected our houses, we heard no signals because the resistance in the more than a hundred meters of wire was too great. I planned to use more batteries, but that night a heavy snowfall brought down the wires. We lost interest in the telegraph, and I soon forgot my rudimentary knowledge of the Morse alphabet.

When I was about ten, the main circuit breaker in our house broke, and an electrician came to install a new switch. I watched him carefully to find out how the switch worked. When he was leaving I asked if I could have the old switch, a glass tube filled with mercury. He gave it to me, and, after breaking it open, I kept the mercury in a bottle in my bedroom and played with it. I took out drops of the shiny metal and scattered them into tiny droplets, which I could join again by pushing them together. When dust collected on the droplets, joining them was more difficult, and inevitably some mercury remained on the floor. When I put a droplet of mercury on a small copper coin and rubbed it between my thumb and index finger, the mercury alloyed with the copper and turned it into a shining silvery piece. I kept some converted coins in my pocket and took them to school to impress my classmates.

Eventually other things became more interesting, and I stopped playing with mercury, but I still don't know whether the exposure to this toxic metal affected my health. It was only years later, when I was a university student, that I learned about chronic mercury poisoning. A fellow student told me about a young chemist who for years suffered from chronic headaches, loss of memory, and fatigue. Finally, his bleeding gums and loosening teeth suggested chronic mercury poisoning. Only then did his co-workers realize that his laboratory room for years had been used for gas analysis and that mercury had been spilled on the wooden floor, lodging in the cracks between boards. When the floor was broken up, they said enough mercury was found under the boards to pay for the repairs.

When I was eleven, I started making tin soldiers, which resulted in further exposure to a toxic metal. After the maids had washed the dinner dishes, I was allowed to use the gas range in the kitchen. The flame was easily regulated, perfect for heating a small crucible in which I melted the metal I used to cast soldiers, as well as pigeons, turkeys, and other farm animals. My source of metal was the foil found around the necks of wine bottles. This lead foil is often alloyed with zinc and is fine for castings. After exhausting the supply of empty wine bottles in our cellar, I found a more plentiful source among the discarded wine and champagne bottles at the Britannia Hotel.

I cleaned up carefully after myself; otherwise I wouldn't be allowed to work in the kitchen. But the work itself was messy. I used a large iron nail to push the slaglike scum that formed on the surface of the hot metal onto the black surface of the stove. At times the two parts of the mold weren't clamped together well enough, and the melted metal flowed out, cooling fast on the cast-iron surface, ready to be used again. Even when the casting was successful, the shapes had uneven edges that I trimmed for reuse. When finished for the day, if there was any metal left in the crucible, I poured small puddles onto the surface of the stove, where it hardened, ready for the next time.

I started painting the soldiers' uniforms red and blue to make them more attractive, and sold them to a nearby candy store for extra pocket money. I soon exhausted my supply of suitable metal and looked for new sources. Visiting a downtown machine shop, I learned that lead was far cheaper than tin and had the advantage of being so soft that I could easily cut it into smaller pieces, suitable for my crucible. Thereafter purchased lead became the chief material for my increased production of soldiers, cannons, animals, or whatever I had molds for and could sell.

All the handling of lead and lead scum undoubtedly contaminated my hands. At that age I paid little attention to washing my hands carefully, and in all likelihood I absorbed more lead than was healthy. Although I seem to have suffered no long-term health effects, I sometimes wonder how much the exposure to lead and mercury affected me.

When I was about ten, I discovered some books on the lowest shelf in Father's library that I was certain I was not supposed to read. However, if I sat quietly on the floor in a corner behind Father's desk, nobody could see me. If somebody entered the room, I could quickly slip the volume back in place and pretend to read a more innocent book I kept at hand.

One of the suspect books was about female physiology and reproduction, probably written in the 1890s. It had no illustrations and was difficult to understand, but I gathered some secrets of life, though the anatomical descriptions were vague. One chapter stressed the importance of a pure life for young women and the dangers of masturbation, rendered with a Swedish word I had never heard. The text promised all sorts of terrible consequences: the ghastly habit would be revealed by pale skin and weakness, sleepless nights, unexplained fevers, perhaps even tuberculosis and death, and if the sinful young woman survived, insanity was inevitable.

Interesting, I thought, and concluded that since the book was about women, such horrors would not apply to males. Aside from my furtive reading and unsuccessful search for more details, I received no information about sex. My mother made one attempt. When I was about thirteen or fourteen she told me to go downtown to see our family physician, Dr. Kindt. She had made an appointment for me to see him in his office, which made no sense to me. I objected that I was perfectly well, and anyway, when he was needed, he always came to our house. Nevertheless, she insisted, and I went.

Dr. Kindt greeted me kindly, invited me into his office, and, sitting across from me at his huge desk, chatted on in a friendly fashion, asking how I spent my day, how school was, and what I usually did after school. He made no medical examination, and when I left I was as puzzled as when I came. Like my parents, he evidently had no idea how to tell a youngster about the facts of life.

AFTER SIXTH GRADE I began three years of middle school, followed by three years in the gymnasium, which prepares students for university. The schools were in the center of town, a twenty-minute walk from home. To get there we had to cross a long bridge over the Nidelven, and in winter when the north wind came sweeping upriver, it was miserably cold.

When the temperature first dropped below freezing, I had to cover my ear on the windy side with my hand to keep it from getting frostbitten. Later, in even colder weather, my ear stayed warm without protection. Much later, when I learned physiology, I understood that the early frost caused vasoconstriction and decreased blood flow to the ear, with the likelihood of frostbite, but once I was acclimatized, even though the weather was more severe, the blood flow in the ear was maintained, keeping it warm.

In middle school I made good grades in science, math, and geography, but in other subjects my grades were at best mediocre. Sometimes I barely squeaked by and was only provisionally admitted to the next grade. My old report cards remind me that I nearly failed German and several other subjects. I hated physical education, and in effort and conduct I had terrible grades, which I certainly deserved.

In drawing class we were graded at the end of spring term on drawings we had made during the year. I didn't have much to show, but a friend who sat next to me was worse off: he hadn't done even one. The teacher started at the top of the alphabet, examined the drawings each boy had made, and gave final grades. My friend was last in the alphabet, so I hurriedly drew a few perspective sketches and passed them to him. My own stilted and elaborate drawings barely got a passing grade, but when the teacher came to my friend, he leaned over. "Well, Torvald," he said, inspecting the work, "you have really improved. Yes, this is good," and he rewarded Torvald with a top grade. Rather than feeling that an injustice had been done, I was immensely pleased to have fooled a teacher I considered stupid.

My feelings for other teachers ranged from scorn to indifference and a reluctant acknowledgment that a few knew their business. My math teacher could hammer some of his subject into even the dullest boy; none failed the final exam. For science and natural history we had Viggo Ullmann, the only teacher I truly admired; he was intelligent, interesting, and always fair. I later learned that the movie actress Liv Ullmann is his granddaughter. I have often wished I could have seen him again, but he was killed in action when German troops invaded Norway in 1940.

By the time I was eleven my interest in animals wasn't limited to what I learned from books. We often spent summers on the coast of southern Sweden, where I saw fishermen clean from their nets starfish and crabs and all sorts of other fascinating animals. I collected a great variety of shells that a friendly curator at the natural history museum in Trondheim later helped me identify. When I learned the scientific names of my shells, it made me feel like a real scientist.

The fall I entered middle school I was twelve. My best friend there, John Stene, was the son of a mathematics teacher who was also an ordained minister; earlier he had been a missionary in Madagascar. John and I started many projects together and frequently stayed for dinner in each other's homes. Although John's family was extremely religious—they said grace before every meal, and playing cards were forbidden—I felt comfortable with them.

Sometimes John and I fished in the river near his house, though we rarely caught anything. We built our own fishing rods, but the best material, split cane, required good tools and was too difficult for us. Instead we used hickory, which we carefully shaved and sanded until we had the smooth, balanced taper required for a good fishing rod.

One day John caught a flounder. I was curious, for why would a fish from the sea swim up the river? And why didn't the fresh water kill it? It wasn't until years later that I learned about Grandfather's successful transplantation of flounders from the sea to an inland lake.

One morning when I was thirteen, Mother surprised me by asking if I would like to try horseback riding. The previous evening she and Father had had dinner with their good friend Dr. Wallem, an art historian and museum director, who had asked if I would be interested in accompanying him when he went riding.

Dr. Wallem, a tall, gray-haired bachelor, was a frequent guest at our house; he was highly cultured, intelligent, a charming conversationalist, and a popular dinner-table partner. Every afternoon he spent an hour or two on horseback and liked to have a young boy get his horses saddled and ready for him. Mother was relieved that I was getting a distraction from school, which I hated and neglected. My earlier fear of horses was long since gone, riding was considered an exclusive activity, and I eagerly accepted.

Dr. Wallem was an excellent teacher and had a thorough understanding of horses. He taught me dressage and eventually jumping, bareback riding, and even a bit of circus performance. One day he made me attempt some of the highly stylized steps of the world-famous Lippizaner horses of the Spanish Riding School in Vienna. One circus trick I learned involved jumping off and immediately back onto a horse while it was cantering. It required a feel for the horse's rhythm and gave me a marvelous sensation of lightness and control—the only time I escaped the feeling of being too tall and lanky.

My growing familiarity with horses was overshadowed by the stimulating presence of Dr. Wallem himself. He was the kind of gentleman who could wear a monocle without looking affected or snobbish. As we rode together, he spoke about history, politics, and economics, about medieval sculpture and religious art, heraldry and family crests, and about Vikings who plundered European cities and brought art treasures home to Norway.

Dr. Wallem was curious about small, everyday matters, such as why most early spring flowers in Norway are yellow, while in August flowers are mostly red or blue. He asked me to explain the physics of radio transmission, which I did, though not to his satisfaction. For a time he tried to teach me French, which he spoke fluently. I listened, but was afraid of sounding foolish and never repeated the French words. When he gave up, I was too embarrassed to let him know how much I wanted him to continue.

The five years I rode with Dr. Wallem were in many ways more important than school. In school, history meant memorizing the names of kings and dates of battles, but Dr. Wallem made history come alive by telling me about real people and real life.

During my last three years with him, I attended the gymnasium, where in preparation for a higher education I could choose between a major emphasis on humanities and languages or one on math and science. I took it for granted that I should choose science. At that time it was commonly believed that girls are unsuited for math and science, and of the thirty students in my class only three were girls.

In addition to math, physics, chemistry, and biology, I was expected to learn three foreign languages: German, English, and French. We also took history, literature, and Norwegian. The Norwegian language as commonly spoken in the cities had developed during a period of political union between Norway and Denmark when the administration was dominated by Danish civil servants. This unbalanced alliance influenced the language; after the union was dissolved in 1814, many Norwegians, motivated by the strong nationalism that swept Europe in the nineteenth century, wanted to rekindle a truer Norwegian language, based on spoken dialects.

Because many Norwegian dialects are difficult to understand and differ greatly from one region to another, the nationalists devised a composite language, known as Landsmål. Norway now has two major languages, the old, traditional Riksmål and the new Landsmål. Different parts of the country can, by local plebiscite, choose one or the other as the primary language. Civil servants are supposed to master both. In school I learned both, but since we always spoke Riksmål at home I chose this as my primary language. In the three-day country-wide final exams we all wrote two essays in our preferred Norwegian language and one essay in the other.

We also mastered a bit of Old Norse, in which the sagas were written a thousand years ago. Old Norse is still spoken in Iceland today, but in both grammar and vocabulary it differs from modern Scandinavian languages more than Chaucerian English does from modern English. Although I value what I learned of the old tongue, I can no longer read it and remember only vestiges.

Such recollection proved helpful some years ago when I was invited for Thanksgiving dinner to the home of a friend, who unexpectedly asked me to say grace. Faced with a table full of guests, but knowing no grace in English and remembering scarcely a word of any in Norwegian, I explained that my words would be in Old Norse, the tongue that prevailed when Christianity was first introduced into my country, implying that my words were appropriate for the occasion. I quoted a few lines of an epic poem that runs in strongly alliterated rhymes and tells how the god Thor awoke from his sleep and found his hammer missing, making him so furious that the earth trembled. Everyone seemed pleased to hear grace in Old Norse and happily proceeded to eat.

WHILE IN GRADE and middle school I frequently had colds or ran a fever, but I was also an expert at feigning sickness. To fake a fever I held the thermometer against a light bulb until it ran up high enough for mother to keep me home, but not so high that she would call the doctor. This kept me at home for a day or two and gave me ample time to read books I liked.

One time I was certain Mother had caught on to my trick when she remained at my bedside, watching me while I took my rectal temperature under the covers. What could I do? I frantically rubbed the thermometer against the sheet, hoping that friction would raise the temperature, but not too much. Mother took the thermometer before I could even glance at it.

"That is peculiar," she said, "very peculiar." The temperature was nearly a degree lower than a few minutes before, but she decided that I still was running enough of a fever to stay home. I didn't remain in bed for long that time, and I never used the trick again.

One reason I could get away with faking sickness had to do with the lack of adequate treatments for infectious diseases, such as antibiotics. In those days, something as simple as a throat infection could cause serious kidney and heart complications. The only measure was to keep a sick child in bed until the fever subsided. My frequent illnesses, some real and some feigned, undoubtedly worried my parents more than I realized.

But my parents had to face more serious emergencies and also a genuine tragedy. In his early teens, Patrick nearly died from a perforated appendix and peritonitis. A few years later he was the victim of a serious shooting accident. He and some friends were shooting crows when a stray bullet from a .22 rifle passed through his left cheekbone, then through his brain, and lodged in the rear of his skull. He remained unconscious for more than a week, and when he awoke the left side of his body was paralyzed. He slowly regained movement, and eventually his recovery was nearly complete. The accident probably influenced his later decision to become a neurologist.

When I was fifteen, tragedy befell my oldest brother, Klas. A top student, a talented athlete and artist, he was frequently presented as a model. Whenever I was scolded for something, I invariably sensed and often heard "Why can't you be like your older brother?" His grade record was the best ever obtained at the Institute of Technology, where his major was chemical engineering, the same as Father's, and Father was immensely pleased and proud of him.

Late one afternoon in early January, during Christmas vacation, I was upstairs reading in my room. Klas was conducting a chemical experiment in a one-room building a dozen meters from the main house. Hearing a noise that sounded like breaking glass, I ran to the window to see flames leaping from the shattered window of the little house. Klas ran out and slammed the door behind him. I shouted down, asking what I could do, and as he ran into the main house, he called, "Turn on the garden hose!"

I ran out to turn on the water, and when Klas didn't return, I ran in again and found him lying unconscious on the floor of the upstairs bathroom. He had no burns, but his face was very red. I rushed downstairs to tell Mother, then outside again to see what I could do about the fire. The surgeon who was our neighbor came by and worked for hours, but all his attempts at resuscitation were useless.

Klas died from carbon monoxide poisoning, which accounted for the bright red color of his face. Carbon monoxide binds firmly to hemoglobin, giving the blood a strong cherry-red color and preventing it from carrying oxygen from the lungs to the tissues. When the fire started, Klas remained too long in the room, trying to extinguish the flames. The oxygen in the small space was quickly consumed, and the fire produced lethal amounts of carbon monoxide. When Klas ran out, it was too late; he was probably dead when I found him in the upstairs bathroom.

The tragedy weighed heavily on everyone. When we sat down to a meal, Father and Mother sat in their usual places, eating little and saying almost nothing. What could we say with Klas' empty place at the table as a constant reminder?

Custom required that black clothes be worn after a death in the family. Before Klas died I had been wearing an old hand-me-down overcoat that was worn and now too small. Father gave Patrick money and asked him to go downtown with me to buy a black winter overcoat. I liked getting a coat I needn't be ashamed of, but I felt enormous guilt at being pleased when I was supposed to be grieving.

The death of a brother I not only admired but also really liked was a heavy blow. It was still Christmas vacation, and I often lay on my bed, reading. There was little else to do, but I felt it inappropriate to read when everybody else was mourning. If someone came into my room, I quickly slipped my book under a pillow.

The death of my parents' oldest and most promising son was a trauma they grieved over for years. After the funeral Klas' name was never uttered again. His death remained a dark burden, an unspoken curse we all remembered but dared not mention.

University Studies

W HEN I WAS in the gymna-
sium I took for granted that I would continue my education and seek a
higher degree. After all, my parents had doctoral degrees, Klas had
planned his doctorate, Patrick was studying medicine, and Astrid was
working for a degree in business administration.

I planned to apply to the Institute of Technology in Trondheim, but
the admission requirements were exceedingly high. Given my mediocre
academic record, my chances of admission were slim unless I applied
to mining and metallurgy, a department with so few applicants that I
might be accepted despite my academic shortcomings. I believed I
could be a competent engineer, although I knew little of what it might
involve; I vaguely imagined that engineers were involved in outdoor
work in exotic countries. I never discussed my future with Father and

got no advice from him, but when I was accepted he seemed pleased that I would stay at home.

The Institute of Technology had one requirement that applied to all students: everyone had to work for twelve months as an ordinary workman in an industrial setting appropriate to the field of study, putting in no less than six months before beginning formal course work.

I started work in the large pyrite mines at Lökken, a couple of hours' travel by bus from Trondheim. A classmate and I were assigned to work with an experienced miner—an old, wiry man who wasn't given to small talk but knew his business from a lifetime in the mines. Our first job was in a horizontal tunnel that was being extended to reach a new ore deposit. We drilled a number of carefully placed holes, about one meter deep, in the end wall of the tunnel.

The drills were similar to the pneumatic impact hammers used to break up street pavement, but much noisier when they went full blast against hard rock in a narrow tunnel. I got used to the noise, as everybody said I would, and nobody mentioned hearing damage. Later, when I worked in a physiological laboratory, I gained a different opinion of "getting used to" pneumatic drills. I found my hearing was impaired above frequencies of about 10,000 hertz, though at that age it should have extended to 18,000 hertz. Luckily, for lower frequencies my hearing has remained normal.

After we had drilled the requisite number of holes, each one was charged with dynamite and a capped fuse. I watched this procedure with apprehension. The dynamite came in cylindrical sticks somewhat larger than a man's thumb. The soft, claylike material consisted mostly of nitroglycerin, with its characteristic bitter-almond smell. We gently pushed several sticks, one by one, into each hole with a wooden stick.

The old miner then took out his pocket knife, cut a suitable length of fuse for each charge, and slipped an explosive cap onto its end. I was astounded the first time I saw him put the cap between his teeth and confidently crimp its edges by biting down hard. Then he inserted the capped end of the fuse into another stick of dynamite and gingerly eased it into the hole. We didn't set off the charges; a special crew came in at night to detonate them, after all daytime workers were out of the mine.

When we returned in the morning, a large heap of loose rock was piled at the end of the tunnel. We broke the largest pieces with a heavy sledgehammer or blasted them again with a small charge, then loaded the rock by hand onto small cars pulled by a miniature electric locomotive along a wobbly spur.

As we extended the tunnel, we laid new track. First, the rocky floor of the tunnel had to be reasonably level. Then we laid the "sleepers," the wooden crosspieces that support the rails and hold them in place. Next we placed steel rails on the sleepers, adjusting the distance between them carefully to avoid derailing. Finally, we fastened the rails to the sleepers with heavy spikes, or "dogs."

The mine gave me experience in a variety of jobs. One day I volunteered to clean the elevator machinery at the lowest level of the mine, about 500 meters beneath the surface. The cleanup was done at night, when only a few special crews remained and the elevators could be kept inoperative.

I already knew that the ambient temperature increases as you go deeper into the earth; even so, I was amazed at how warm it was. When I worked at higher levels, I had to wear a heavy sweater under my coverall; at the lowest level I wore nothing under it. The hot steam I used to blast old grease away from cables and gears intensified the heat, and I understood why the job was unpopular. I didn't volunteer for it again.

After several months in the mine I was transferred to the mechanical shop above ground where machinery and pneumatic drills were repaired. Although I liked it there, I preferred working in the mine. However, at this time I also started wondering if I would rather return to my old interest in biology.

Before I began engineering, I had considered medicine. At that time Norway had only one university with a faculty of medicine; it had open admission, so I would have been admitted in spite of my mediocre grade record. I was reluctant, however, because Patrick was a student there. It was bad enough to be the son of a distinguished professor whom everybody seemed to know; even worse was always being compared with an older brother.

I believe it was also the desire to escape being compared with other family members that made me decide to leave engineering, although the reason wasn't clear to me then. All I knew was that Father would disapprove and be disappointed; somehow he seemed to hope that I could replace the son he had lost and continue the family tradition in engineering.

I went around for days working out in my mind how I would tell Father; finally I got up my courage and went to see him. I looked at his stern face and blurted out, "Father, I want to study zoology!"

It was clear from his frown that he didn't like the idea, but he did not discuss my reasons or his objections. With his unspoken disapproval and reluctant permission, I could enter the University of Oslo.

Today I would act differently when talking with Father. I would start by telling him I needed advice. I would explain that I had gained valuable experience from engineering and had thought carefully about my future. I would ask his opinion about natural science. Then I would mention my interest in zoology, an area he was already familiar with. How different it would be, not only for me but also for him.

Anyone who had finished the gymnasium could matriculate at the university. Thus, after half a year as a miner, I entered the University of Oslo in January 1934 to study zoology. The transition was abrupt. At the gymnasium, subjects and textbooks were prescribed, and every hour of the day was scheduled. No such structure existed at the university. I was not required to register for courses or to attend lectures, and nobody seemed to care whether I did or not.

In the beginning Patrick gave me some helpful advice. He was in Oslo as a medical student, and we lived in the same private dormitory, where a rather austere breakfast and supper were provided daily. "You must find out what suits you here," he told me. "Some lecturers are much better than others. Your study plan is up to you." After this I had little contact with Patrick. We were not particularly close, and at meals in the dormitory he usually sat with his fellow medical school students; I was left to fend for myself.

For an advanced degree, I had to choose a major subject and two minor subjects. Zoology majors, for example, might choose botany and paleontology as minors, or perhaps chemistry or physics, depending on their interests and inclinations. Before advancing to the major, every student had to pass a final comprehensive exam in the minor subjects, usually after two years of lectures without any intermediate tests. The zoology lectures covered subjects such as comparative anatomy, taxonomy, genetics, histology, embryology, invertebrate zoology, and animal geography, with all these subjects included in the final comprehensive examinations.

As a science student I was required to take a one-semester course in calculus and a one-year course in philosophy and psychology. My performance in calculus was miserable. During the oral exam, the old professor, who was retiring at the end of the term, gave up and sighed, "Since you are my last student, I will give you a pass." I have often regretted that he didn't give me the failure I deserved: knowing more math would have served me well later in life.

My performance in philosophy was better. I went to few lectures, but I read the recommended books, one in philosophy, one in formal logic, and one in psychology. I absorbed enough to pass the course, though without distinction.

The author of the psychology text was a young professor, a trained psychoanalyst who had studied in Vienna. Parts of the book dealt with functions such as hearing and seeing, subjects that I now call sensory physiology. Then came the classical aspects of human psychology and the professor's specialty, the unconscious mind, which I found fascinating. At that time psychoanalysis was a highly controversial subject, though one rarely mentioned in newspapers and magazines. Freud was not yet a household word, and because in the public mind psychoanalysis implied sexuality, the subject was not considered "nice." Reading the text, I was amazed to learn that a person could have thoughts and desires that remain unconscious and that unrecognized fantasies could be revealed through associations and analysis of dreams. I was keenly interested and might have changed to medicine except for my resolution not to follow in Patrick's footsteps.

I was also determined to learn to read German fluently because of its importance in scientific literature. The zoology professor gave me permission to read German zoology books, sitting in a corner of the laboratory room reserved for advanced zoology students. I started on the first page of a book on laboratory techniques and forced myself to read every word to the very last page.

In the long run this book proved very useful. It explained how to make glass pipettes, how to build a thermostat, how to sharpen a microtome knife, how to preserve animals, and lots of other practical information. I applied myself to this task with more determination than to anything else I did at the time, and at the end of my first term at the university, I could read German effortlessly.

During my years in Oslo I was also privileged to know Kristine Bonnevie, the head of the Zoology Department. Her clear and well-organized lectures covered a wide range of subjects during the two-year lecturing cycle.

One day a lovely young woman, Hilde Wegener, appeared in Professor Bonnevie's house. Hilde came from the University of Graz in Austria. Her father, a professor, had died three years earlier during an expedition in Greenland, and Hilde had come to spend a year as a houseguest of Professor Bonnevie.

We soon learned that Hilde's father, Alfred Wegener, was the eminent geophysicist who had originated the hypothesis of continental drift. He thought that Greenland was slowly moving away from the European continent and wanted to determine the rate of the drift. His theories were rejected and belittled, if mentioned at all. However, my memory of Hilde, who was kind and beautiful and melted the heart of every young zoology student, was more than sufficient to maintain my unflagging faith in the concept of continental drift. It is now well established that Wegener was right, although the movement of continents is better known today as plate tectonics.

During my three years in Oslo I attended all zoology lectures as well as lectures in other subjects I found interesting, such as geology, paleontology, and botany. This enjoyable academic freedom helped me acquire a broad scientific background, although it did not contribute to rapid completion of a degree.

The person who influenced me most during those years was a friend named Iacob Sömme, a man about ten years my senior whom I admired for his keen mind and friendly charm. He was interested in fish biology and especially in the brown trout, the most important sports fish in Norway. He had no academic position and lived on a variety of stipends and small research grants.

One day Iacob suggested that I should focus my work on physiology and study in Copenhagen with August Krogh, Scandinavia's most famous physiologist. I barely knew what physiology was and had never heard the name Krogh. Iacob explained that before I tackled physiology, I had to know more chemistry. I argued that I already knew all the chemistry I needed from reading chemistry books I found at home. "If you already know so much," Iacob countered, "it will be easy for you to take the exam and get it over with." I had sense enough to heed his advice and plan for a minor in chemistry.

The requirements for the minor included a one-semester laboratory course in analytical chemistry, which I was permitted to take at the Institute of Technology in Trondheim. The engineering students there were much more career oriented than my student friends in Oslo. One of them had several years of industrial experience and an excellent knowledge of practical chemistry; he taught me not only chemistry but also a great deal about life in the real world, both as an analytical chemist and as an industrial employee.

The chemistry students also arranged some wild parties. Because of the exorbitant taxation of commercial alcoholic beverages, we utilized other sources. Alcohol was used in the analytical method to determine the amount of potassium in a sample. A carefully rationed amount of a few hundred cubic centimeters of alcohol was issued for this difficult but popular procedure. The instructor undoubtedly knew that the laboratory alcohol would be redistilled and consumed, but he also knew that it is useful to economize with chemicals and to gain experience in distillation.

After my stint in the Trondheim lab, I returned to Oslo for the spring term to study for the final chemistry examinations. This was the only term during my years in Oslo when I worked truly efficiently. I lived in a rented room not far from the science campus. Every morning I woke up around seven and read for a couple of hours. At ten I walked to the university for the first lecture, and after that I read again. In midafternoon, after eating a hot meal in the university cafeteria, I walked back to my room and continued studying. In the evening I ate a spartan meal of bread and butter in my room and read again, lying on my bed until I fell asleep about ten. Because I made no effort to study long into the night, I woke every morning without an alarm, rested and ready to read. By maintaining this routine for the entire term I learned a great deal of chemistry in a fairly short time.

The chemistry examinations began with two days of written tests, with one day devoted to inorganic chemistry and the other to organic chemistry. The written inorganic exam went well because I had used an advanced German text that stressed principles rather than memorization. The organic exam was worse; I had to write for three hours about anthraquinones, and I didn't have three hours' worth of knowledge.

Then came two days of orals. My examiner was the physical chemist Odd Hassel, later a Nobel Prize winner, who was obviously delighted that I understood electron shells and, without glancing at the big chart on the wall, knew the periodic table both horizontally and vertically. In contrast, my performance in the organic oral was about as unsatisfactory as possible; I was repeatedly stumped when asked to name various methods of organic synthesis. Nonetheless, I passed with a fairly good overall grade; evidently my knowledge of inorganic chemistry made up for my deficiencies.

In the summer of 1935, after passing the chemistry exam, I joined Iacob Sömme on the mountain plateau in central Norway where he was studying the reproductive biology of trout. He taught me about fertilization of trout eggs, about the role of pH in the water, and about acid runoff from peat bogs. To my great sorrow I learned after World War II that he had been executed by the German Gestapo because of his activities in the Norwegian underground resistance.

Another older friend, Per Höst, joined us during the summer in the mountains. In addition to being a scientist, Per was a talented animal photographer who wrote several excellent books about animal life. Like

Iacob, Per had a flair for pinpointing important problems. He had studied seal biology in the Arctic, and when he joined us in the mountains, he was investigating the reproductive biology of ptarmigan. He taught me some basic field skills, such as how to find ptarmigan nests and determine the population density. We also studied the role of lemmings and voles as a food source for predatory birds, so I learned firsthand about predator-prey dynamics.

It was well known that the populations of rodents fluctuate wildly from year to year, although the reasons weren't well understood. Per had observed that the mountain buzzard and other birds of prey raise a greater number of young in years when the lemmings and other rodents are numerous and provide ample food. He was now eager to find out how in early summer, long before an explosion in rodent population was evident, the birds seemed to know to lay more eggs in anticipation of a rich food supply later in the season.

That summer I understood the value of personal contacts with scientists and their daily work. Listening to Per and Iacob discuss their ideas was very different from the formal lectures of professors whom I otherwise rarely saw, and I began to see how exceedingly important personal interaction is for the development of stimulating new ideas. That summer I undertook my own small research project on the population densities of voles and lemmings, and found that the results varied according to how I baited the traps.

It was Iacob who stimulated me to undertake a study that resulted in my first substantial research publication. I addressed a simple question: Are brown trout and Arctic char, the two most important freshwater fish in Norway, competitors for the same food resources?

I began my research in the summer of 1936 when Father and I went fishing at a mountain lake, Grönningen, near the Swedish border. This fairly inaccessible and undisturbed lake was ideal for the study Iacob had suggested. For three summers I weighed and measured every trout and char we caught, examined its stomach contents, and took samples of its scales. Fish scales have growth rings, somewhat like tree rings, and can be used to determine age. By examining the scales, I determined the growth rate of the fish and the age composition of the population. By analyzing the stomach contents, I could determine food preferences. What the fish ate ranged from crustacean plankton to insect larvae, snails, and an occasional small fish. Overall, trout fed more on insect larvae and plankton, while char consumed more of bottom-living organisms such as snails, but there was considerable overlap between the two species.

Over the three summers I amassed information from 723 fish, and in 1939 I published the results. I had learned how to collect information, how to tabulate results, how to prepare graphs, and how to express myself reasonably well in English. I felt great satisfaction when the paper was published.

Those summers weren't devoted only to science. We enjoyed the fishing, and Father taught me how to smoke the trout, the largest of which weighed more than three kilograms, or nearly seven pounds. With Father's long experience and insight into food processing, the result was as good as the finest smoked salmon.

During my winters at the university I saw Father whenever he came to Oslo on official business. Although he always invited me to dinner at a good restaurant and treated me as an adult, I was never entirely at ease in his company.

I ALWAYS WENT home to Trondheim for Christmas vacations and for at least part of the summer. I didn't see much of my old classmates from school, but I read a great deal, mostly books that had no bearing on my studies. In retrospect I wish I had visited my old mentor, Dr. Wallem, who had always been fascinating to listen to during our long horseback rides, but it never occurred to me that he would care and even be pleased if I went to see him at his museum. I was in my early twenties and perceived adults as people who were fundamentally uninterested in idle youngsters like me.

In contrast, Mother had come to view me as a full-fledged adult, someone in whom she could confide. Now in her late fifties, she told me about her unhappy life, much of it wasted. She complained bitterly about how inconsiderate Father had been and what a terrible burden it had been to be married to him for so many years.

During one of these long talks she also told me how angry Father was when she became pregnant for the fourth time, in effect telling me I was an unwanted child. This didn't particularly shock me; rather, I felt she was trusting me with confidential information. Mother also spoke about what I assumed was the abortion of a fifth pregnancy, performed in Oslo by one of Father's old friends, a well-known gynecologist, and of how terrible she felt on the train back to Trondheim. She said she deeply regretted that she had not left Father years earlier and taken the children with her.

No doubt Arrhenius, her professor in Sweden, considered her an exceptionally talented young physicist before she married and moved to Norway. Now she was embittered because Father had prevented her from continuing her research into the behavior of electric discharges in gases at low pressure.

In addition to her duties with home and children, she helped Father in his laboratory, where she served as a highly skilled technician. Over the years her name had appeared on many joint publications; but only on her very first publication, her doctoral work, had she been the sole or even the first-named author.

Mother felt that her collaboration with Father reflected only his interests, which concerned vitamins A and D. "My real talents in math and physics were wasted," she said. She also maintained that, had Father listened to her suggestions, the results would have been far more significant. Vitamins were still mysterious substances, known only through their effects, and the clarification of their chemical structures might have brought them a Nobel Prize. Indeed, years later Albert Szent-Györgyi received the Nobel Prize for clarifying the chemistry of vitamin C, and Henrik Dam for his discovery of vitamin K.

Mother was an exceptionally sincere and honest person, and because she undoubtedly fully believed everything she told me, so did I. I sympathized with her resentment, for she had always been closer to me than Father, whom I saw as distant and stern. I now perceived him as very selfish and much less intelligent than Mother. I did not yet realize that people can perceive the same facts in different ways.

It is easy to understand Mother's need to talk. She was alone after her children left home, living in a large, empty house with a husband for whom she apparently had little affection. Obviously, it would have been better not to involve one of her children in her unhappiness, but it was not yet considered socially acceptable to consult a psychiatrist for personal problems.

How real were her complaints? True, she had worked hard, both at home and in Father's laboratory. She had kept an open house with many guests and parties, and she had been frugal, making do with less household money than she needed. She saw Father as essentially her opposite, selfish and never willing to sacrifice his own needs and comfort.

Our long conversations made a deep impression on me and undoubtedly affected my own life. When I later married an intelligent and ambitious woman, I did all I could to help her career, fearing I might do to her what I perceived my father had done to my mother. Only years later did I come to understand the serious challenges to a marriage between professionals in related fields, and as I understood my own feelings better I began to see my parents in a more realistic light. As a result, my attitude toward my father shifted. I began to see him as a man with scientific curiosity and competence, well-intentioned but perhaps inept at expressing his feelings.

Back at the university I continued my zoology studies. One of my fellow students, Thor Heyerdahl, had an important though indirect influence on my life. Heyerdahl was very much an outdoor person, well built and athletic. If it snowed, he might set out on skis for several days in the mountains with his husky and a sleeping bag made of reindeer skin. Once I asked him if it was uncomfortable to sleep outdoors in winter. "Yes," he answered, "because you must sleep with your boots on your stomach to keep them from freezing stiff."

During his third term at the university, Heyerdahl decided that formal zoology wasn't what he wanted, and on a stormy day in October he announced his decision to leave. The term had started with a study of amoebas and paramecia, living animals we could see moving about under a microscope. On that particular day we had moved up the animal hierarchy to coelenterates. Trays filled with preserved sea anemones and corals were set out for study when Heyerdahl walked in, rain-soaked and dripping wet. He took one look at the animals, and as he turned around and walked out he exclaimed, "Damn it, these are flowers. I quit!"

Over the next weeks we saw little of him. By Christmas he had married his girlfriend and was on his way to the Marquesas Islands in the Pacific. The couple spent a year on these beautiful islands, living off the countryside. Heyerdahl would later become famous for the *Kon Tiki* expedition, which carried him and his crew on a raft across the Pacific Ocean, and for other adventurous expeditions that continue to stir the public imagination, making his name a household word.

Before Heyerdahl left for the Pacific, he sold me three of his textbooks. One of these became the most influential book I have ever owned. It was the original edition, in German, of Richard Hesse's *Ani-*

mal Geography on an Ecological Basis. In fact there is little classical animal geography in it. Instead, it discusses the distribution of animals in relation to their physiological characteristics. It covers life in the high mountains and in the depths of the ocean. It examines the importance of high temperatures and lack of water for life in deserts, of the salts in the ocean for osmotic problems, and of the lack of salt for life in fresh water. Hesse's book brought to my attention questions that have occupied much of my scientific life: What physiological characteristics permit some animals to live in environments that to others are hostile and uninhabitable?

Most of the half dozen students majoring in zoology enjoyed a good time. We had frequent parties, taking advantage of our easy access to alcohol in the zoology laboratory. We diluted the alcohol to a concentration of 35 or 40 percent and added a few drops of caraway oil to make a reasonably good aquavit. Of course it was illegal, but it was difficult for the professor to know exactly how much alcohol escaped from the intended scientific use.

One fall I spent several weeks with a friend, Edward Barth, at the university marine station. We were alone, working harder than usual and using a fair amount of alcohol while for a change drinking none. When the director of the station came on his weekly visit, he displayed ill-disguised suspicion when Edward asked for more alcohol, but he issued the requested several liters, mumbling disapproval. This made me suspicious. After the director left, I poured a few drops of the alcohol on a clean glass plate to let it evaporate. It left a thin layer of a white powder, which turned bright red when I added a drop of sodium hydroxide. This showed that the director had added phenolphthalein, a strong laxative, presumably to cure us of drinking lab alcohol.

We were indignant that our innocence was in doubt, conveniently forgetting our customary use of lab alcohol. I suggested that we should pour the entire supply down the drain and ask for more. The next time the director came, Edward innocently asked for several more liters. The professor looked puzzled but issued the alcohol without comment, and without phenolphthalein.

One of our most memorable parties was the result of Per Höst's interest in badgers. Badgers mate in the fall and give birth in early summer. Per wanted to know if the fertilized egg undergoes delayed implantation and starts development only after the animals become active in early spring after sleeping through the winter in their deep dens; otherwise, the fetus must go through a period of arrested development. To answer the question, Per had to examine the ovaries of a female in the fall; accordingly, he set out on a badger hunt with some friends.

Per returned with a female badger and dramatic tales of hunting with a dachshund. "With its short legs," Per said, "the dog follows the badger into the den and drives it out. You have to be careful, though, because an angry badger may rush out and attack you, biting your legs without letting go until it hears the sound of crushing bones." The hunters had protected themselves by wearing high boots stuffed full of pieces of charcoal; when the badger bites and hears the crunching sound, it lets go.

What good is a dead badger after its ovaries are removed? In the fall, badgers are good to eat, having gorged themselves on acorns in preparation for the long winter sleep. Per's badger looked made to order for a feast, and Per persuaded the supervisor of the students' cafeteria to roast the badger like a suckling pig.

That evening, without the professor's knowledge, we had a tremendous feast in the main teaching lab on the top floor, featuring a stuffed badger from the zoology collection as the centerpiece on the table. Per's badger, beautifully roasted, was brought up from the kitchen in the basement, young and tender and tasting like roast pork. Half a dozen of us ate and drank until we were ready to burst.

DURING MY time in Oslo, music became an important part of my nonacademic life. The main university building in the center of the city housed a large assembly hall with excellent acoustics that was often used for concerts. A number of free tickets were usually distributed to students, and whenever I had a chance I asked for one. One concert remains etched in my mind because the performer became part of the music. I sat near the stage, listening to Josephine Baker's lovely voice. She was right in front of me, in a long red dress that clung to her slim body as she moved with the music. I absorbed her beauty with all my senses. At age twenty, I saw, felt, and understood for the first time that sex is more than mere copulation.

My interest in music gave me another opportunity when the National Theater had a guest performance by the Royal Swedish Opera. Students served as extras when needed in crowded scenes. Three operas played to sold-out houses. In one, my role was to look grim and ugly in a crowd of medieval peasants. The second, a modern opera, I don't remember at all. The third opera was Bizet's *Carmen*. I appeared several times in different costumes, ending up in the last act as a picador, wearing glorious clothing with heavy gold embroidery. Carmen was played by a dark-eyed and slim beauty, more Spanish in looks than Swedish. She was lively and charming and sang with a wonderful voice.

The next time I heard *Carmen* was in Leipzig. This time Carmen was a tall, forceful, blond Wagnerian soprano with a tremendous bosom. She thundered across the stage like a galloping horse, hurling her dramatic performance at the audience, a parody of the vivacious Carmen I had seen in Oslo.

This was in 1935, when Father traveled to several major European universities to study chemistry education and had invited me to go with him. By then much of central Europe was affected by Hitler's political gains. After World War I, the Versailles treaty prohibited Germany from rearming, a prohibition openly flouted by Hitler. One day when I was at the Berlin Zoo, a formation of warplanes thundered by, right over my head. It was obvious and well known that Germany was preparing for the war that started in 1939 with the invasion of Poland.

One friend of Father's was the Jewish biochemist Karl Oppenheimer, who had been banned from his institute in Berlin when Hitler took over Germany. We went to see him in a small, cramped room where he lived alone. He told us about a long monograph on physiological chemistry he had recently completed. It was to be published in Holland because no German publisher dared take it. I asked him how he could write a comprehensive monograph when he was not allowed in the university libraries. He explained that while he still could, he had spent a year reading the pertinent literature, and the following year he had written the book. I was deeply impressed, not only by his prodigious memory, but also by the desperate dedication that went into completing his work under such conditions.

In Czechoslovakia the dislike of everything German was evident. Most Czechs could speak German, but they preferred not to. If we spoke to a taxi driver in German, he would pretend not to understand.

I found a way around this problem by speaking in English. Drivers usually didn't know English and invariably asked if I knew German; then we could communicate freely.

When in Copenhagen on our way home, I heeded my friend Iacob Sömme's advice about studying physiology. I wanted to meet August Krogh, the physiologist who in 1920 had been awarded the Nobel Prize for his research on the role of the capillaries in supplying oxygen to the body tissues, studies that brought him world fame. I phoned Krogh's institute and asked if I could visit, and when I arrived I was shown into Krogh's laboratory. The professor was in the middle of an experiment, so I had a chance to observe him while I waited. He was slim with narrow, stooping shoulders, thin blond hair, a gray mustache, and a pointed beard. Then Krogh greeted me with a big smile that bared his large and protruding yellow teeth. He spoke so rapidly that I had trouble understanding while he explained how he determined the permeability of frog skin to water by placing a frog in what looked like glass trousers. One trouser leg contained ordinary water and the other heavy water as a tracer substance. He could then follow the penetration of water through the skin of one leg by measuring the heavy water that appeared in the other. I left, impressed by the friendly reception and certain I would like to come back to study under this eminent scientist.

THE TRAVEL through Europe had taken me away from the university for two months, but it didn't matter much; students were free to come and go as they wanted. I had, however, missed the lectures on insects in the invertebrate zoology course. Traditionally, the zoology exams did not include questions in entomology, and the students considered this an unwritten law, so I made no attempt to catch up on what I had missed.

In 1938 I registered for the final zoology exam. I managed reasonably well in the written exams. The orals went even better until it came to the identification of a large tray of animal specimens. I quickly named the marine animals, which I knew well from field trips and from a summer spent on a research vessel off the Arctic coast of Norway. I carefully avoided a large insect with red spots on its extended wings, hoping the professor might not notice. However, he pointed to it, asking what it was. I tried ducking the question with a vague hint that it might be some kind of butterfly.

"Thank you; that is enough," the professor said, ending the exam. The insect was a large grasshopper, mounted with extended wings. I was embarrassed, but I passed.

Coming to Grips with Physiology

FATHER APPROVED of my inter-
est in physiology and wanted to be of help, so he wrote to Professor
Krogh to ask if I might be permitted to work in his laboratory as a student. The eminent Nobel laureate replied in the affirmative, and in
September 1937, a few weeks before my twenty-second birthday, I
enrolled at the University of Copenhagen.

When I arrived in Denmark I had difficulty understanding what
people were saying, which was especially troublesome when scientific
problems were discussed. Although written Danish is similar to Norwegian, it takes months before a Norwegian can readily understand a
rapid conversation in Danish. The numbers are particularly difficult,
and numbers are important in the daily life of a scientist. Danes count
to 100 in multiples of 20 instead of 10, as in English or Norwegian. For
example, 60 in Danish is *tres,* an abbreviation for three times 20. Even

worse, 50 goes by the seemingly illogical term *halv-tres*. The ones are spoken before the tens, so that twenty-one becomes one-and-twenty. Imagine, for example, that in English, instead of saying fifty-six, you would say six-and-half-sixty, and you get the point.

An Austrian physicist, Otto Frisch, who worked at Niels Bohr's famous Physics Institute and later helped develop the atomic bomb, complained that although he spoke Danish fluently, he found it almost impossible to understand. Even after five years in Denmark he often was unable to understand a group of Danes when they chatted among themselves. He had a theory that the Danish language first developed among fishermen shouting over the breakers at the coast, where consonants are inaudible. He maintained that Danish contains no consonants, only a thick soup of vowels interrupted by an occasional glottal stop.

I gradually learned more about the contributions to science of August Krogh and his wife, Marie. She had a degree in medicine but did not practice, except to see a few patients referred to her for metabolic studies. She also had a moderate research program, concerned with the metabolic effects of the thyroid hormone.

Long before I came to Copenhagen, the Kroghs had collaborated in studies that revolutionized existing ideas about how oxygen moves from the lung air into the blood. Early in the century two of the world's best-known physiologists, J. S. Haldane in England and Christian Bohr (Niels Bohr's father) in Denmark, had maintained that oxygen was transported from lung air to blood by a special transport mechanism, needed to move oxygen molecules across the lung membrane. Likewise, it was believed that carbon dioxide was discharged into lung air through a similar active transport process.

Working in Bohr's laboratory, Krogh devised methods to show that all gas movements between lung air and blood take place by diffusion; that is, oxygen moves passively, without the need for any special transport mechanism. The Kroghs published their studies in 1910 in seven consecutive papers that forever demolished the theory of active secretion. It was a bold move for an unknown young scientist to contest two of the world's most distinguished physiologists. Krogh made many other contributions to human physiology that now are standard textbook knowledge. He was, for example, a pioneer in the study of the effects of exercise on humans and remained a leader in this field. He was also a technological genius, designing new methods of analysis to solve problems that formerly seemed unapproachable.

While still an assistant in Christian Bohr's laboratory, Krogh devised a method to examine the effect of carbon dioxide on oxygen binding in blood. A study published in 1904 and authored by Bohr, K. Hasselbalch, and Krogh used Krogh's method to show that carbon dioxide causes a reduction in the affinity of hemoglobin for oxygen, an effect now universally known as the Bohr effect.

I have wondered whether Krogh's contribution was greater than is suggested by the appearance of his name as the last among the authors. Not only did the paper clearly bear the mark of his genius, but in his book *The Comparative Physiology of Respiratory Mechanisms* Krogh never used the term "Bohr effect." Instead, his graph illustrating the carbon dioxide effect was simply labeled "Krogh," without reference to the original paper. Today one cannot know how Krogh felt about the effect being permanently connected with Bohr's name.

Although Krogh's many discoveries are basic to human physiology as we know it today, his contributions to animal physiology are even more varied. Over the years he solved numerous problems of insect respiration, worked on the swim bladder of fish and on diving insects, and much later, after his retirement, studied insect flight and the physiology of migrating locusts, problems that had intrigued him all his life.

When I came to Krogh's laboratory in 1937 he was concerned with aquatic animals and how they regulate the concentration of salts in their bodies. Freshwater organisms are in constant danger of losing essential salts to the surrounding nearly salt-free fresh water, but marine animals are surrounded by a medium with ample amounts of salts.

Marine invertebrates have high salt concentrations in their bodies, similar to the concentration in sea water. However, the relative amounts of various salts in their blood differ, and they must therefore be able to regulate their content of sodium, potassium, magnesium, and other minerals. For fish the situation is different; to survive, marine fish must maintain salt concentrations in their bodies about one-third those of sea water. In the 1930s the mechanisms necessary to regulate these concentrations were a prime focus of Krogh's studies.

One of the first problems Krogh suggested I tackle was a comparative study of osmotic regulation in freshwater crayfish and marine shore crabs. The shore crab can penetrate far into the Baltic Sea, where the salt concentration is very low, but not all the way into fresh water, so what is the difference between them?

I found that in brackish water the crab can maintain higher blood concentrations, but in very dilute sea water or in fresh water it cannot and dies. However, another marine crab, the Chinese mitten crab, is able to maintain its salt concentrations even in fresh water; thus it can invade the large European rivers, where it causes great damage by tunneling into dams and river banks. Nevertheless, the mitten crab must return to the sea to breed.

When I was working on these problems, I was reminded of the flounders that my grandfather in 1903 had released in a freshwater lake. Like the mitten crab, they could survive in fresh water but needed the sea to reproduce.

The Physiological Institutes in Copenhagen were located near the university hospital. One large building, known as the Rockefeller Institute, housed four departments: Medical Physiology, Biochemistry, Human Exercise, and finally August Krogh's domain, Zoophysiology, in which important research in human physiology was conducted. When the institute was built, the university provided a residence for each department head as well as for the second in command—the equivalent of a senior lecturer.

Krogh had a spacious villa directly connected to his laboratory so he could attend to his experiments at any time, day or night. Every morning at seven he went to his laboratory and, although nobody would dare disturb the otherwise friendly professor, locked the door. At nine he returned to his residence to dress and eat breakfast, and he was back in the institute when work officially started at ten.

When I first arrived, the secretary had arranged for me to rent a room from the senior lecturer in biochemistry, Henrik Dam. Dr. Dam, who discovered vitamin K and was awarded the Nobel Prize for this discovery in 1943, had no children and no need for a maid, so I rented one of two empty maids' rooms in the attic. Dam was a small, dry man whom I saw only once a month to pay the rent. He was affable but not talkative; when he did talk, he spoke very fast in a low voice and was even harder to understand than most Danes.

The other maid's room was rented to Mr. Knappeis, a Jewish refugee from Hamburg. In the early and mid-1930s a constant stream of Jewish refugees came to Denmark, helped by private citizens as well as the authorities. Because resources were limited, many continued on to other countries, often England or the United States. Knappeis, however, remained in Denmark, working as a highly skilled technician for the well-known muscle physiologist Fritz Buchthal. Quiet and modest, Knappeis was a small, thin man who looked as if starvation had stunted his growth during the miserable years after World War I. I believe that he received little more than a token salary; Denmark was still suffering from the worldwide depression of the 1930s and could offer only meager support for the many refugees who were in need.

I usually ate my dinner in a nurses' dining room a few blocks away, but Knappeis seemed unable to afford even that modest expense. He prepared his dinner on an electric hot plate in the hallway outside our rooms; then he put up a small table in his room and set it with a clean tablecloth, a plate, and silverware. When the meal was ready, he carried it to the table, whereupon he walked out again to return to a dinner that was prepared and served for him.

This civilized routine impressed me deeply. At that age I certainly wouldn't have made so much out of eating alone. Many years later, when I was divorced and lived by myself, I remembered Knappeis. Then I understood how important it is for one's self-respect to eat like a civilized person, rather than grabbing food on the run. I adopted Knappeis' routine, and when eating alone, I always set a table before sitting down to dinner. I have remained grateful for what I learned from him.

Professor Krogh had four children. The oldest, Erik, was a few years older than I and was studying medicine. We were already friends when I arrived in Copenhagen because the preceding summer Father had invited him to join us in the mountains for fishing. Erik and I had hiked and fished and got along marvelously well. After he completed his studies, Erik was appointed to the Anatomy Department at the University of Aarhus. He was a gifted researcher, original but erratic, and at times suffering from manic depression, as did several relatives.

One day I asked Erik's youngest sister, Bodil, out for an evening. The next older, Agnes, worked as a technician in the laboratory and was engaged to a Swedish student, Christer Wernstedt. The oldest, Ellen, was a practicing dentist and had left home.

Bodil was nineteen, with wavy blond hair and blue-gray eyes. Her pointed nose and receding chin gave her a likeness to a painting of a Renaissance lady. She had just started dental school but still lived with her parents next door to the institute, so I saw her frequently; often we didn't go out at all but remained in the house or in her room.

In the late spring I asked my parents if she could spend the summer with us in the mountains. They agreed, and in mid-June 1938 Bodil and I traveled by train north through Sweden and Finland, then by bus into Norway, and finally by ship along the coast to Trondheim. The mountain cabin where we would spend several weeks had only two bedrooms, and Mother asked if it wouldn't be more proper if she and Bodil shared a room. She seemed greatly relieved when I said that it wasn't necessary.

Bodil and I spent increasing amounts of time together, and everybody assumed our relationship was permanent. We often took long bicycle trips, and at times we went to Krogh's country house for weekends, a bicycle ride of several hours.

I was treated like a member of the Krogh family. When I suffered from a gastric ulcer, Marie Krogh was concerned and prepared bland food for me, the standard treatment for ulcer patients. Despite having frequent pains and neglecting my research, I managed to finish a substantial paper describing my fish studies, writing laboriously in English with diligent use of a dictionary. August Krogh generously helped me by correcting my manuscript as it progressed.

When Mrs. Krogh invited guests to a formal dinner party, I was always included. A frequent guest was Dr. H. C. Hagedorn, a physician who in the early 1920s had worked with Krogh to start large-scale insulin production. Hagedorn had devised a novel method to determine

blood sugar, known as the Hagedorn-Jensen method, faster than old methods and requiring only a few drops of blood.

At one of these dinner parties I met the Hungarian-born chemist George de Hevesy, who had discovered a new chemical element that he graciously named Hafnium, from the old Latin name of Copenhagen. Later, when my research involved the use of radioisotopes, I came to know him better.

Several times the physicist Niels Bohr was there. He was in his early fifties, compact and stocky. Though an excellent soccer player as a student, he now had a stooping posture, as if his huge head were weighing him down. Bohr always spoke softly, and one had to strain to hear him. When he lectured, he was seemingly unaware of speaking to an audience. His secretary was always seated at a small table in front of him, taking down every word.

I remember Bohr best for a discussion of inheritance. After dinner one evening the conversation turned to the question of inheritance of intellectual traits, a possibility that Bohr adamantly denied. I suggested that musical ability certainly is inherited; the famous soprano Kirsten Flagstad, for example, came from generations of musically talented individuals. Bohr was unconvinced and maintained that his own sense of music was the result of his mother's singing to him when he was small. "Musicality depends only on childhood experience and how one is brought up," he insisted.

When someone mentioned the similarities between identical twins, Bohr countered by telling a joke. "Yes," he said, "and in America they have experimented by separating twins at birth and bringing them up differently, and they still come out alike. They have sent them to different schools, even one to Harvard and the other to Yale, and yet they turn out alike!" This ended the discussion, everyone as unconvinced as when we started.

Krogh had an impressive command of English and a singular interest in the language. He was especially fond of Kipling, had read virtually everything Kipling wrote, and could quote long passages even though he claimed to have a poor memory. I saw convincing testimony of Krogh's command of English when he published a monograph on osmotic regulation with Cambridge University Press. When the proofs arrived, he asked Christer Wernstedt, his future son-in-law, and me to help him read the proofs. I expected that a meticulous English publisher would suggest many changes, but no more than half a dozen points were raised in the whole book. Christer and I agreed that this was impressive for someone who had spent only brief periods in an English-speaking country.

I learned a great deal from reading the proofs for Krogh's osmoregulation book. One problem that especially intrigued me was how marine birds survive with no fresh water to drink. In search of an answer to this question, I developed methods I could use under primitive conditions, and at the end of the spring term in 1939, after securing permission from the Norwegian authorities to capture birds, I set off for the coast of northern Norway to try to solve the problem. This journey was the first of what would become a long series of field studies around the world, ranging from the Sahara Desert to the Amazon River, seeking answers to problems of how animals survive in hostile environments.

In mid-June 1939 I arrived at Röst, a small island in northern Norway, off the Lofoten chain and facing the Arctic Ocean. Millions of auks, puffins, and gulls nest on vertical cliffs that rise out of the ocean beyond Röst. The birds seek their food at sea, and except for rainfall there is no fresh water. Do they drink sea water? I wanted to find out.

It was already known how whales and seals can manage. If they drink sea water, the extra salt is excreted by the kidneys. Whale kidneys are powerful and can produce urine more concentrated than sea water. Although no one knows whether whales and seals actually drink sea water, they could readily eliminate the excess salt.

A female whale that nurses her young needs additional water to produce milk, but the situation is less problematic than one might think. Cow's milk is relatively dilute and contains only about 4 percent fat, so producing it takes a great deal of water. Whale's milk, in contrast, may

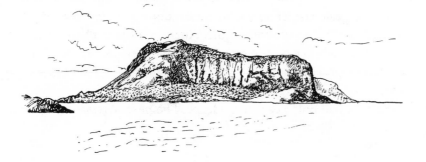

contain as much as 50 percent fat, a fact first reported by my father. Whales thus have the ability to transfer concentrated nutrients to their offspring, using much less water than other mammals do.

Humans have less powerful kidneys than seals and whales. A human castaway at sea who drinks sea water merely hastens the approach of death because the kidneys are unable to excrete the excess salts. Birds seemingly are worse off; their kidneys are even less efficient than humans' in eliminating salts. The problem was to find out if birds get sufficient water in their food, or if they drink sea water and somehow are able to excrete the salts.

My first step was to determine how much salt there was in the birds' food. If I found substantially more salt in the birds' urine than there was in the food, this would suggest that the bird drank sea water. Because most of the salt in sea water is ordinary sodium chloride, I could get the information I needed with a simplified analytical method I had developed.

If the birds drank sea water, it would also show up in another way, based on the fact that roughly one-tenth of the salts in sea water are magnesium salts. Most of the birds' food has a very low magnesium content; if I found a large amount of magnesium in the feces or urine of a bird, it would indicate that the bird drank sea water. I had developed a method for measuring magnesium, but it was complex and in the end wasn't of much use to me.

When I arrived at Röst with my equipment, a local merchant let me use an empty old boathouse, where I worked sitting on the rough wooden floor. It was primitive but adequate. No more than a score of families lived on the island, and nearly all of them made their living from the sea. Tons of codfish hung to dry from long racks of horizontal wooden poles, to be exported as stock fish to the Mediterranean countries.

During the summer I examined skuas, auks, puffins, and kittiwake gulls. The salt concentration in their guts was invariably low and showed no evidence that any of them drank sea water. I also examined five seals shot by a local fisherman; the results were similar. Neither the salt nor magnesium content in the seals' stomachs and intestines suggested that they had drunk sea water.

So far I had only negative evidence. The next step was to find out what happens when a bird actually swallows sea water. I captured a few kittiwake gulls and caged them in empty orange crates. They greedily devoured the fish I fed them. Fish doesn't have a high salt content, so I gave one of the birds an ample volume of sea water by stomach tube. If the kidneys excreted the salts, there should be a high salt concentration in the urine.

The bird produced copious volumes of urine, but to my amazement the urine had little salt in it. I repeated the experiment with other birds, and again the urine was nearly salt-free. Wondering if my analytical methods were wrong, I tested every step with solutions of known salt content; my methods were 100 percent correct.

No matter how much sea water I gave the birds, little salt appeared in the urine. Could it be that the birds retained the salts? If so, the salt concentration in the blood should increase. But my analysis of their blood showed no elevated salt levels. Where was the salt going? I knew it had entered the body, yet I couldn't find it in the urine or in the blood. It seemed that the salt had simply disappeared.

At the end of the summer I returned to Copenhagen, disappointed that I had found no solution to the original problem. I was anxious to talk to Dr. P. B. Rehberg, a prominent renal physiologist, who usually gave young scientists excellent advice. However, he said little, and I felt that he perhaps thought I hadn't done a very good job; he didn't even look at my meticulously kept data books. In desperation I suggested that if the salt doesn't come out the rear of the bird, it must somehow come out the front. Rehberg didn't comment.

I had one hypothesis that I didn't mention to Rehberg: perhaps the gastric epithelium of marine birds, which normally secretes hydrochloric acid to aid in digestion, is specialized to excrete a more concentrated solution of chloride ions. Excess sodium, I imagined, could be bound as insoluble urate and excreted in the urine.

I wanted to tackle this problem again, but the war intervened, and then other projects took all my time. Not until eighteen years later, in 1957, did I return to the study of marine birds. As I had suggested to Rehberg, a salt load is indeed eliminated from the front end of the bird, as a salty fluid dripping from its beak. Then I understood why I hadn't noticed the phenomenon when I was on Röst. The primitive conditions where I worked, the orange crates and the rough wooden floor, made it difficult to see drops of fluid the birds shook from their beaks. That summer in Norway, I thought the few drops I noticed were no more than a little sea water regurgitated by the bird.

Except for the failure of the bird study, my summer on Röst was delightful. Summer weather on the coast of Norway is often cold and rainy, but that year the good weather seemed to last forever. I spent almost every night outdoors and rarely returned until early morning. Being that far north of the Arctic Circle, with the sun shining all night, I needed little sleep, often only two or three hours, but was always ready to work the next morning.

One day one of my local friends said that two German students had pitched a tent nearby and might need my help to communicate. When I went to see them I found two tanned, healthy looking men with well-trained, muscular bodies who looked more mature than most students. Rather than accepting my offer, they firmly declined any help and clearly discouraged further conversation. Unlike most students, who make do with little equipment, they carried an impressive array of cameras, more than any tourists I have ever seen. It seemed obvious to me that these "students" were German naval officers on a photographic mission. The site they had selected was also telling; the outer islands of Lofoten control the main approach to the port of Narvik, the major export site for high-quality iron ore from Sweden.

With the threat of war hanging over Europe, I reported the apparent spy activities to a visiting police official. He was helpless, for there was no law against German visitors photographing our countryside or studying our important harbors. Yet the Germans clearly posed a threat to national security. When German forces invaded Norway less than a year later, a major battle for control of Narvik took place in this area.

CHAPTER 5

The War Years

By THE END of the summer of 1939, the situation in Europe was tense. Hitler's troops had occupied Austria; Czechoslovakia had conceded its western areas to Germany in exchange for a "peace" agreement, which was violated only weeks later when Hitler's troops invaded the country. In Spain Franco had established his fascist regime, and Italy invaded Albania. In Scandinavia we did not perceive the events in the Mediterranean area as an immediate threat; we had become accustomed to the lengthy Spanish civil war, and Albania seemed remote. However, the German aggression in central Europe scared us.

Then Germany signed a nonaggression pact with the Soviet Union, freeing Hitler's troops for other military action. On September 1, 1939, they swept into Poland. Britain and France declared war on Germany but were unable to provide any effective military assistance. As German troops advanced toward Warsaw, Soviet troops invaded from the east, and one more country fell into the hands of the dictatorships.

We were uncertain as to what might happen next. Bodil and I had talked about marrying and eventually moving to Oslo. As we spent much of our spare time together, being married would scarcely change our lives. On September 20 we were married in a brief ceremony at Copenhagen's City Hall, followed by a wedding dinner for family and close friends. After a brief visit to Stockholm, we returned to Copenhagen, where Bodil continued her dental studies and I, my work in the laboratory.

Not long thereafter, Scandinavians were shaken by the news that the Soviet Union was making territorial demands on Finland. The Finns rejected the demands, and in November Soviet planes bombed Helsinki. Now we felt directly threatened. The Finns put up heroic resistance, and with the arrival of winter the mechanized Soviet troops made poor progress in the snow-covered landscape. The other Scandinavian countries helped the Finns as best they could, and Sweden contributed as much as a neutral country dared, sending every sort of supplies short of troops that might be needed at home. But the Finns could not for long sustain a war against the overwhelming Soviet forces, and in March 1940 they signed an armistice agreement yielding substantial territories and permitting the Soviets to establish a naval base on Finnish territory.

For the spring term of 1940 I went to Oslo to take the final exams for a magister degree, somewhat like a Ph.D. in the United States. Bodil remained in Denmark to continue her work in dental school. I didn't much like being alone in Oslo, but I studied hard and spent little time on diversions. During Easter vacation we had a welcome break in our hectic exam schedules when Bodil joined me for a week of skiing in the mountains; then she returned to Denmark.

One evening I went to a public lecture given by Niels Bohr at the Norwegian Academy of Science. He discussed recent research on nuclear fission and expressed doubt that a chain reaction could release the tremendous amounts of energy that theoretically were possible, a statement probably intended to mislead any German intelligence agents in the audience.

German forces attacked Denmark and Norway without warning on April 9, 1940. German troops, which had been hiding in two coal freighters in the Copenhagen harbor, stormed the city early in the morning while mechanized land forces crossed the border from Germany into Jutland. The Danish government surrendered, and Denmark was occupied; resistance would have been suicidal.

In Norway the Germans met heroic resistance, losing two cruisers and many destroyers, submarines, and supply ships to the coastal defense. Such heavy losses slowed the German advance, but the Norwegian land forces were unable to sustain effective resistance. The Norwegian cabinet and the king escaped to Britain to establish a government in exile. As I had anticipated during my summer on Röst, Narvik with its important harbor was taken by German forces. Norway remained an occupied country until the end of the war.

I was in Oslo during the invasion. Fortunately, there was no fighting in the city itself, but we didn't know what to expect. How would the Germans punish us for their heavy losses? Would there be any food? Where could we be safe, if anywhere?

The telephone system didn't work, and there was no public transportation. To contact my brother, Patrick, on the other side of the city, I set out on an hour's walk. The university was closed, and I gave no thought to exams.

A friend and I volunteered at the University Psychiatric Hospital for whatever work was needed. Many patients had been evacuated, and we took the night watch for those more seriously ill who remained. We stayed there for several weeks, feeling some small measure of security within its walls.

One day rumors that the Germans were preparing to bomb Oslo spread through the city, giving rise to general panic. Only months earlier the Russians had bombed Helsinki, and many thought that the Germans would similarly attack the civilian population of Oslo. Many tried to get out of the city, but the roads were jammed, and besides, where could anyone be safe? In the end we were spared.

One month after invading Denmark and Norway, German forces entered Holland, Belgium, and France. I was still in Oslo, but with the university closed I had no reason to remain in Norway. I applied to the

German occupation forces for permission to travel to Denmark, and to my surprise I was allowed to return to Copenhagen. I arrived on May 19, forty days after the invasion of our countries.

The situation in Denmark was far better than in Norway and other occupied countries. Denmark produced food, and even though huge quantities were shipped to Germany, enough remained for the civilian population. As a result, the Danes never suffered starvation as did people in Holland, where nearly all food was shipped off to Germany. In addition, Germany treated Denmark as a model of "friendly" occupation, in contrast to the general terrorism in other occupied countries.

The occupation forces imposed a total blackout on news from the rest of the world. Listening to radio transmissions from the Allied countries was strictly prohibited. Danish newspapers, subject to German censorship, reluctantly printed useless propaganda news. We nevertheless stayed informed through underground news distributed on typewritten and mimeographed sheets. As the war continued, the growing resistance movement in Denmark caused increasing problems for the Germans.

In Copenhagen the university was open, and I was granted permission to enroll for degree work. With my osmoregulation study as my thesis, I passed the exams and in February 1941 received from the University of Copenhagen the degree I had intended to obtain in Norway.

Bodil received her dental degree several months later. We had now lived under German occupation for more than a year and had begun to feel the impact of war shortages on our lives. Our bicycles were essential for going to work, but it was impossible to buy a set of new tires. We kept our old, worn-out ones going by tying pieces of rags around places where the tires threatened to explode.

Bodil began practicing dentistry and was soon appointed an instructor at the Copenhagen School of Dentistry, initially in general dentistry and later in surgery. Although pregnant, she continued teaching and doing surgery, riding her bicycle to work every day, until our daughter Astrid was born, on October 16, 1941. For about a week immediately afterward, we engaged a trained nurse, Miss Mortensen, who had been highly recommended by Bodil's brother, Erik, as the best possible person to help out and teach us how to handle our new parenting duties.

At that time the authorities on child care considered it best for both babies and mothers if strict feeding schedules were maintained, and Miss Mortensen was a disciplinarian who knew what was needed. I shudder at the thought of what we were told—feed the baby exactly every four hours, don't give in to crying, don't pick up the baby because it rewards bad habits, and so on. There is so much I regret and would do differently today, but both Bodil and I wanted the best for our children, and how could inexperienced parents dare disregard what knowledgeable authorities advocated?

Although food was rationed, we could usually buy meat, eggs, and milk, and we never went hungry. Letters from my parents revealed a different picture for Norway. All correspondence was censored for anything that could be interpreted as intelligence or as criticism of the occupation regime, but occasionally some information got through. One summer, for example, Mother described how happy she and Father were for a ration of two eggs, the first since Christmas; in the middle of a sentence she slipped in the words "and both were rotten."

One of the few things I could do for my parents was to send them food packages. An export license was granted only for small amounts to be sent once per month to close family members. I sent food my parents hadn't seen for years—cheese, liver sausage, and dry salami. I also tried to send eggs, but although I packed them carefully, too many arrived broken. I continued sending food parcels throughout and after the war; occasionally I also sent a package to my sister, Astrid, in Stockholm, where the food situation was difficult, although not as critical as in Norway.

In Denmark the worst shortage was fuel. No oil was imported, and only small amounts of coal reached the country from Germany. Consequently, there was no gasoline for cars and trucks. A few essential vehicles were equipped with an ingenious device based on the combustion of wood. A large steel cylinder, like a tall wood-burning stove, was attached to the side of the vehicle. The cylinder was filled with small blocks of wood, and a fire was started from below. When the air supply was reduced, the combustion formed gases rich in carbon monoxide, which were conveyed to the modified engine and powered the vehicle. Though highly inefficient, the device helped maintain a minimal amount of slow-moving traffic.

The lack of fuel was especially difficult because we suffered three winters of record cold, with temperatures remaining near −20°C, or zero Fahrenheit, for weeks on end. Electricity was severely rationed, with a tenfold penalty levied for exceeding the tiny allowance. When Bodil and I studied in the evening, we both read by the light of one single 25-watt bulb. Gas for cooking was severely rationed, and hot-water heaters were prohibited. Only rarely could we heat a little water for washing, and cold showers were unthinkable when the rooms were icy cold. Families with small children had an extra gas ration, enough to heat a small amount of water on the kitchen stove to wash the baby, but Bodil washed Astrid's diapers by hand in cold tap water. Bodil developed a painful case of mastitis, but there was no effective treatment, and her discomfort was intensified by the penetrating cold in our rooms.

We were allowed one puny cake of soap a month, of which 50 percent was clay, added as an extender. We managed well, however, because I made our own soap from surplus cooking fat. In a big kettle I heated the fat with a solution of sodium hydroxide until the hot mixture suddenly turned to soap. Then I added kitchen salt, which caused the pure soap to separate and float to the top to form a thick, ivory-colored layer while impurities remained in the salty brine below. Having enough soap made life easier for us than for most people.

One sad but perhaps inevitable consequence of such rationing was the emergence of various scams. I recall two incidents in particular. In one, a homeowner had bought a load of peat on the black market to heat his house and had it delivered and dumped in his basement. Peat isn't good fuel, but it is better than nothing. Still, if too wet, it won't burn, and it was therefore forbidden to sell peat with more than 50 percent water content. The homeowner's pile melted and formed a large brown puddle; what had looked like solid peat was mostly solid ice. Of course, the seller was long gone.

In another case, the papers reported that a well-dressed gentleman had appeared in various offices and offered black-market cigars at an outrageous price. The cigars were neatly packed in boxes with the brand name Zostera printed in beautiful letters on the lid. People who were desperate for tobacco greedily bought them by the boxful without even tasting them. The smokers quickly regretted their purchase; there was no tobacco in the cigars, only seaweed. They complained to the authorities, and the gentleman was brought to court but not convicted; he had never said that he was selling tobacco and had made no claims for quality. To the contrary, he had honestly named the "cigars" Zostera, the scientific name for the seaweed they contained.

As THE WAR progressed, the underground resistance movement increased its activities. Rail lines were blown up to impede the transportation of food to Germany, and factories forced to produce goods for the Germans were sabotaged. These activities were directed by a shadow government whose leaders remained unknown until after the war, when we learned that Dr. Rehberg had been one of them.

At the institute we suspected that Dr. Rehberg was active in the underground. He was rarely at the institute, but one day he showed up and asked me to help him. To be certain we were not overheard, he took me into the elevator, and while it was moving he hurriedly said he needed 1,000 kroner in cash. He showed up a few days later, and we again got into the elevator, where I handed him the money I had withdrawn, almost all I had in my account. It was to pay for the printing of an underground paper carrying forbidden news about Allied progress.

The less a person knew, the safer everybody was. The rules were simple: Don't ask questions; don't know names.

Despite diligent German surveillance, the resistance movement maintained secret escape routes to Sweden. This underground transport system saved many Allied pilots who were shot down óver Denmark, and it was the salvation of Danish Jews once it became known that the Germans planned to round them up. Within a few days nearly all Danish Jews were smuggled to Sweden, where they remained safe until the end of the war.

The Danish police refused to collaborate in the German roundup of Jews. They also consistently abstained from any action to stop the extensive sabotage activities. Eventually the Germans retaliated. In 1944 the entire Danish police force was arrested and deported to a concentration camp in Germany.

One of those saved was Niels Bohr, who would be targeted not only because he was a top physicist but also because his mother was Jewish. One dark night in the autumn of 1943, Bohr and his family were smuggled across the sound to Sweden in a small fishing boat. From Sweden he was taken to England in a two-engine Mosquito bomber. As the plane flew over occupied Norway and ascended to a higher altitude, above the reach of the German anti-aircraft guns, Bohr lost consciousness. Because the headphones didn't fit his very large head, he hadn't heard the pilot instruct him to turn on the oxygen supply. The pilot was worried about Bohr's silence but had no choice but to remain at altitude until he had crossed Norway. When the plane descended over the North Sea, Bohr regained consciousness. After the war, Bohr's son Aage corroborated this story.

From England, Bohr sailed to the United States, where he helped develop the first atom bomb at Los Alamos. He was accompanied by Aage, a physicist who later succeeded his father at the University of Copenhagen. To ensure secrecy, both men were given false names: Nicholas and Jim Baker. The physicist Otto Frisch, who also worked at Los Alamos, once asked Bohr how he got along with his false name and whether he remembered to sign himself as Nicholas Baker rather than Niels Bohr. "What is the difference?" Bohr replied. "My signature is illegible anyway."

Otto Frisch also tells a story about Bohr's arrival in the United States. Secret agents who carried the luggage rushed Bohr and his son from the ship to a taxi and then to their hotel, where father and son were registered under their assumed names. After hustling the two to their room, the agents expressed great relief that the identity of the important visitors had not been revealed. Then one of them noticed the name Niels Bohr in large black letters on his suitcase!

Although the university laboratories in Copenhagen remained open, few of us remained focused on our research. Not only were supplies and equipment scarce, but also nearly everybody was in some way involved in the underground resistance. If nothing else, someone was always needed to carry messages or distribute the illegal news sheets that sprang up everywhere.

During this period I started working with an instrument maker at the institute, thinking that shop experience would be useful in my work. He suggested I gain experience by building the mechanical parts of a microscope, adding lenses later. As my project progressed I became familiar with the proper use of a drill press, lathe, milling machine, and other shop equipment. I learned to harden and temper tool steel to make the specialized tools I needed. That shop experience has proved extremely helpful to me throughout my career. Researchers often need to overcome mechanical problems or design new devices, and shop experience makes it easier to solve problems as they arise. When the help of a mechanic is needed, they can often suggest feasible designs, while scientists who lack shop experience often ask a mechanic for devices that are impractical and sometimes outright impossible.

The instrument maker, Mr. Person, started making parts for rapid-firing handguns, called tommy guns, that were secretly produced for the underground army for possible combat with the Germans when the war drew to a close. I sometimes helped him cut the tough Martin steel by hand with old, dull hacksaw blades; no new blades could be obtained anywhere. The pieces were machined to the correct shape and tolerances and then moved to another illegal shop, where they were case-hardened before final assembly with parts fabricated elsewhere.

The wailing of the air-raid sirens became familiar, and we knew the difference between the deep rumbling of German planes and the higher-pitched, comforting hum of the Allied planes. We never went to the air-raid shelter, for we knew that the Allied planes aimed at targets such as industrial sites especially important to the Germans.

Most bombing raids took place at night, but one occurred in full daylight, aimed at the Gestapo headquarters in the center of Copenhagen. To prevent attacks on the building, the Gestapo had converted the top floor into a prison to hold important Danish leaders. Among the prisoners was Rehberg, although the Germans never were able to pinpoint his role in the underground movement.

During this particular bombing raid, a series of British Mosquitos came in at a very low altitude, just above the rooftops, and dropped bombs that went in at an angle through the side of the building at the level of the lowest floors. A tragic aspect of the otherwise successful raid was that a few bombs fell on a nearby school, where several children lost their lives.

The prisoners on the top floor of the steel and concrete building were shaken but unharmed. One of them grabbed the keys from a stunned German guard and unlocked the other cells. Most of the prisoners, Rehberg among them, found their way out and escaped. Rehberg remained in hiding until the end of the war.

One prominent scientist working in Copenhagen was the Hungarian chemist George de Hevesy, who was attached to Niels Bohr's institute. Hevesy has probably influenced modern physiological and biochemical science more than any other person. He introduced radioisotope techniques into biological research; in fact, most of our current understanding of cellular and molecular biology depends on the use of radioactive tracers.

When Hitler took power in Germany, Hevesy came to Copenhagen. He spoke half a dozen European languages fluently and with the same accent: English, French, and Danish all sounded like Hungarian. Hevesy often rode his bicycle to Krogh's institute and greeted us with a friendly question he immediately answered himself: "Is everything all right? Yes!" (in Danish, "Gaar det godt? Ja!").

Contrary to common belief, Hevesy had been experimenting with isotopes long before the artificial radioactive isotopes were produced in the mid-1930s. From 1912 to 1914 he worked in Manchester with the physicist Ernest Rutherford, developing techniques to use radium D as an indicator for the presence of lead. This technique permitted Hevesy to make otherwise impossible measurements, such as determining the minute solubility of lead compounds considered totally insoluble. Even more amazing, he found that one solid metal can diffuse into another.

Many years ago I heard the tale of Hevesy's first radioactive tracer experiment, one that never made it into the scientific literature. While in Manchester he lived in a boardinghouse where the meals were less than appealing. There, as in boardinghouses everywhere, the tenants surmised that hash was made from leftovers and plate scrapings from the previous day.

To confirm his suspicions, one night Hevesy surreptitiously added a small amount of polonium to the food he left on his plate. The next day he brought to dinner an electroscope, consisting of a glass case in which a light strip of thin gold foil stands out horizontally when the instrument carries an electric charge, but hangs down limply when the charge is lost. Hevesy rubbed his pocket comb against his woolen vest and transferred the static charge to the electroscope. The gold leaf stood straight out until the expected platter of hash appeared on the table; at that moment the gold leaf dropped to the discharged position, indicating the presence of ionizing radiation. Hevesy is said to have moved to a different boardinghouse. If the story is true, this may well have been the world's first radioactive tracer experiment.

Hevesy was always initiating biological experiments. Because Bohr's institute was dedicated to theoretical physics and had no facilities for animal work, Hevesy sought collaborators at other institutes, riding around on his bicycle, often with a vial of radioactive phosphate in his pocket, asking colleagues to conduct an experiment for him. He might, for example, ask to have a fertile egg injected with radioactive phosphate to see how phosphate was distributed in the chicken when it hatched. Among his collaborators were August Krogh and Hans Ussing, as well as several researchers at the university hospital. He also worked at the Carlsberg Laboratory with the plant physiologist Carsten Olsen, who studied phosphate transport in plants. Indeed, there were few research institutes in greater Copenhagen where Hevesy did not have experiments in progress.

Colleagues said that Hevesy was clumsy in the laboratory, and there were many stories about broken equipment and other mishaps. Once he broke a Kjeldahl flask containing hot concentrated sulfuric acid that splashed over his arm, burning him severely. He went to the emergency ward of the nearby university hospital, where he was already known from earlier visits, and returned, pale and obviously in pain, with his arm bandaged. When his secretary suggested that he go home and rest, Hevesy replied, "Do you think it will hurt less if I go home?"

Along with other prominent scientists, Hevesy was compelled to leave Denmark during the war, and in 1943 he appeared with his family in Stockholm, where he was welcomed at the Karolinska Institute. He became a Swedish citizen and remained in Sweden until his death in 1966 at age eighty.

After several years at Krogh's institute I wanted to spend some time at the Carlsberg Laboratory to learn some ingenious microchemical methods developed by Dr. K. Linderström-Lang, which I thought would be useful in physiological research. With permission from Linderström-Lang I started working at the laboratory, which adjoins the famous Carlsberg Breweries on the southern outskirts of Copenhagen.

Linderström-Lang was perhaps the most talented person I have ever known: an outstanding chemist, an accomplished violinist, a fine singer, and a much admired painter. He projected an aura of innocence but was in no way childish, just fun-loving. Taller than average and slightly stooped, he had straight blond hair that always dropped over his forehead, and he brushed it back with a characteristic quick movement of his hand.

The Carlsberg Laboratory was established in 1876 by the owner of the Carlsberg Breweries, J. C. Jacobsen, who was both a visionary and a true intellectual. He kept in contact with leading scientists and introduced new methods in his brewery as they were developed, such as the refrigerating machines invented by Karl Linde. He understood the significance of Pasteur's work on microbes and why bacterial contamination turned beer into a foul fluid, fit only to pour down the drain.

The Carlsberg Breweries expanded rapidly under the leadership of this farsighted brewer. Backed by a healthy company, Jacobsen established the Carlsberg Laboratory for research, stipulating that no discovery of theoretical or practical importance be kept secret. He then established the Carlsberg Foundation, with a board of trustees composed of outstanding Danish scientists named by the Royal Danish Academy. More than a century later the Carlsberg Foundation is still the foremost supporter of research in Denmark.

Until 1938 the director of the Carlsberg Laboratory was S. P. L. Sörensen, whose name is inseparably linked with the concept of pH. It was Sörensen who first described the acidity, or hydrogen-ion concentration, of a solution according to a logarithmic scale, using the symbol pH. Many scientists believed that Sörensen deserved a Nobel Prize for his introduction of the pH concept, but others thought him ineligible because his work did not include an actual scientific discovery. Still, the concept of pH is an integral part of the daily life of scientists and affects every aspect of chemistry, physiology, biochemistry, and medicine.

Linderström-Lang, who was always addressed as Lang by everyone in the laboratory, succeeded Sörensen as director. He developed ingenious methods for determining minute quantities of protein-, carbohydrate-, and fat-splitting enzymes in tissue samples of less than one milligram, improving levels of detection a thousandfold. He was able to determine the oxygen consumption of a single amoeba, at the time a measurement so minute that it seemed beyond imagination, and he designed a novel method to determine the minute density changes in a protein during enzymatic cleavage. He was at ease with thermodynamics and mathematics, which were important in his work with the structure of proteins, then poorly understood.

Although the research at the Carlsberg Laboratory was at the cutting edge, the atmosphere was collegial and lighthearted. At lunchtime we would sit together at a long table with Lang at one end and bottles of beer in the center, eating our slim wartime sandwiches. Lang contributed greatly to the warm and friendly spirit. Typical of his spontaneity and love of fun was his invention of the "Carlsberg Rocket." Removing the small piece of metal foil from inside a bottle cap, he folded it around the head of a safety match. He then balanced the match carefully at an angle in the mouth of a bottle and held a second burning match under its head until the first suddenly took off in some unpredictable direction.

At Carlsberg I concentrated on developing new methods of fat analysis, aimed at characterizing the fats in tissue samples no larger than one milligram. This work had little to do with other activities in the laboratory but a lot to do with a problem I was curious about—namely, how whales can dive and remain submerged for hours, presumably without any oxygen. Whales have a large amount of blubber with a high content of unsaturated fatty acids; that is, they are rich in double bonds. The fact that blubber is richly supplied with blood vessels might mean that blubber is metabolically highly active. Could it be that, during diving, the double bonds of the fatty acids could serve in the place of oxygen as a hydrogen acceptor, that is, take the place of oxygen in intermediary metabolism?

My hypothesis was that the high degree of unsaturation of whale blubber might explain some of the physiological mysteries of diving. I had no way to experiment with whales, but I tested the idea on earthworms and tadpoles, which I thought of as tolerant to lack of oxygen. I deprived them of a normal oxygen supply but found no indication that the double bonds of their fatty acids were affected. As often happens in science, it was a dead-end idea; I abandoned all thought of whale blubber and turned to other problems.

My next project was more rewarding. One day Carsten Olsen, a plant physiologist at Carlsberg, casually mentioned that spinach has a high content of oxalic acid, which to me presented an obvious question: Since oxalic acid is poisonous, how could spinach be such a desirable food for humans? I decided to study the effect of oxalic acid and spinach on laboratory animals. I felt I had come up with my first good and original idea for a scientific study.

I knew from elementary chemistry that oxalic acid combines with calcium to form an insoluble compound, calcium oxalate, which is harmless. However, because oxalic acid binds preferentially to calcium, the result should be calcium deficiency, which in turn should result in weakened bones and poor calcification of developing teeth. More serious consequences arise if there is more oxalic acid in the food than can be bound as calcium oxalate. The excess oxalic acid then lowers the calcium concentration in the blood, which in turn causes muscle spasms and nerve malfunction, typical of oxalic acid poisoning. Although oxalic acid is excreted in the urine, crystals of calcium oxalate form in the kidney, eventually producing kidney stones.

Wanting to understand the physiological effects of oxalic acid ingestion, I started feeding experiments with white rats. I asked Bodil to participate in the study, not only because she was a careful and competent collaborator but also because her work in dentistry made her especially interested in the effects of calcium deficiency on the teeth. Rat teeth, as opposed to human teeth, continue to grow during the life of the animal and therefore provide a continuous record of tooth calcification.

Despite the war, I obtained some newly weaned rats from the university hospital. Food was rationed, so I applied for and was granted special rations for rats. Because no edible oil was available during the war, our rats ate real butter mixed with flour and dry milk powder, a diet that made us jealous because our own supply of butter was severely restricted. To calculate how much spinach to add to the rats' regular diet, I devised micromethods for determining the exact amount of oxalic acid in spinach and the level of calcium in the rats' blood.

As the weeks went by, it was obvious that we were on to something important. The rats that ate spinach grew more slowly. Their serum calcium was abnormally low, less than half the normal level for rats.

Their bones were poorly calcified and had a greatly reduced calcium content. Parts of the continuously growing teeth that formed after the experiment started were poorly calcified, with the calcified dentin about a quarter of the normal thickness.

Could some component of the powdered spinach other than oxalic acid have these effects? This question was easily answered. We added calcium carbonate to the spinach-containing diet and fed it to a new group of rats. The extra calcium alleviated all the deficiency symptoms. To make certain it was the oxalic acid that was harmful, we replaced the spinach with an equivalent amount of pure sodium oxalate. The effects of spinach were exactly duplicated, making it obvious that oxalic acid was indeed the harmful factor.

Finally, we examined the effect of spinach on the kidneys. First, we added to the diet an amount of spinach to match the calcium in the food, but to yield no excess oxalic acid. A microscopic examination confirmed our expectation: we found no effect on the kidneys. We then gave the rats the same proportion of spinach and regular food as before, but in two separate meals. One was the regular food without spinach, and the other was a carefully calculated amount of spinach that we made tempting by mixing with butter. In each twenty-four-hour period the rats thus ate exactly the same proportion of basic feed and spinach as the rats in the preceding experiment. After several weeks their renal tubules were nearly clogged with a solid precipitate of calcium oxalate, exactly as expected.

We decided that Bodil should report our results at a meeting of the Danish Dental Society. To our surprise, the newspapers found out about the lecture and reported, as sensational front-page news, the discovery that spinach is poisonous.

The news, reported in late spring, just when the first crop of spinach was brought to market, caused some panic. People were afraid to eat spinach, and we received threatening telephone calls and letters from gardeners unable to sell their crop. Fortunately, the excitement gradually subsided as others emphasized the nutritional value of spinach.

What bearing, if any, did our experiments have on human nutrition? First, persons who have a tendency to form kidney stones should not eat spinach or rhubarb, a long known fact. For people who are otherwise well-nourished, reasonable amounts of rhubarb or spinach probably do no harm. However, because humans, like rats, form calcium oxalate, it is prudent to avoid a daily intake of spinach over long periods.

After the newspaper reports came out, a pediatrician friend told me about a case our studies helped resolve. A boy nearly two years old had not yet learned to walk and could not sit up straight. His bones were poorly calcified, and he was calcium deficient. The underlying cause became obvious when it was learned that the boy's father, a gardener, had grown spinach year-round in a heated greenhouse (an illegal use of fuel during wartime). Thinking he was giving his child the best possible diet, he mixed fresh spinach into the baby's every meal. When I heard about this case, I felt that despite the economic loss to the gardeners, the sensational publicity had done some good.

With the spinach study completed, I thought of problems more promising than whale blubber, for I still had to complete a doctoral degree. Armed with the new methods I had developed at Carlsberg, I returned to Krogh's institute to study how fats are digested and absorbed in the intestine.

The subject was controversial. It was well known that digestive juices contain enzymes that split fats into glycerol and fatty acids. In the nineteenth century the German physiologist Eduard Pflüger had suggested that fatty acids, together with sodium bicarbonate from the pancreas, form water-soluble soaps that are absorbed. Other investigators had different theories, maintaining that soaps cannot exist under the slightly acidic conditions in the intestine. Because bile is necessary for the normal digestion of fat and acts as a detergent, fat would be emulsified, somewhat like the droplets of butterfat in cream. These minute droplets could then be absorbed, and there would be no need for the formation of soaps.

The absorption of sugar was better understood and suggested a possible approach. During absorption, glucose is bound to a phosphate group as glucose monophosphate. I wanted to see if, during absorption, fatty acids likewise bind to phosphate to form phospholipids. I planned to use radioactive phosphate, which was produced in small amounts at Bohr's institute.

I did find a greater formation of phospholipids during fat absorption, an increase apparently not caused merely by a heightened metabolic activity in the intestinal wall, for when glucose was absorbed, there was no such increase in phospholipids. Therefore, my results indicated that fatty acids may indeed be transported as phospholipids, although my experiments did not conclusively prove this.

I conducted this study during the last years of the war, when it was impossible to buy chemicals. Still, we were careful with the little we had, and I managed well by working with micromethods.

During these years Bodil was also engaged in research, studying the solubility of tooth substance in relation to the composition of saliva, information important for understanding how dental caries develops. She needed novel analytical methods, and because I had become a reasonably good glassblower, I made the specialized glassware she needed for the microanalysis of calcium. I also taught her to make and calibrate her own micropipettes.

One step in her analysis required that a precipitate of calcium oxalate be heated. We had no electric furnace that could reach the desired high temperature, so I built one from scrounged materials, using a refractory porcelain cylinder and a few meters of Nichrome resistance wire I begged from a machinist friend. I found a jar of asbestos powder to use as insulation after I mounted the porcelain cylinder inside an old tin can. My calculation of the length of wire needed to reach the desired temperature turned out to be correct, and everything was ready for the project.

Bodil worked with tremendous energy and completed her research project while teaching at the School of Dentistry. She continued working through a second pregnancy and was back at work shortly after our son, Bent, was born, on March 19, 1944.

MARIE KROGH died in 1943 after a lengthy illness, never having told her children that she had breast cancer. We later learned she had consulted a prominent surgeon but had declined an operation, perhaps because she suffered from diabetes, another condition she kept secret for years. During Bodil's mother's long illness, she, Astrid, and I had moved back to Krogh's residence, where Bodil took charge of the household.

War was still a major factor in our lives. One evening in June 1944,

Krogh's telephone rang. By chance I was near the telephone and answered. The caller identified himself. "This is Professor Aubeck of the Technological Institute," he said. "May I speak to Professor Krogh?" Krogh was in his laboratory, less than twenty meters away, but on an impulse I replied, "No, the professor is not at home."

Assuming that Krogh's telephone was bugged, I walked the few steps to the laboratory and used its phone to call Professor Aubeck and ask if he had called Krogh a few moments earlier. "No," was his surprised reply. "I haven't made any call." Now the situation was obvious. The Germans had started retaliating against the increasing Danish sabotage by gunning down well-known public figures, and since Niels Bohr's escape the previous year, Krogh was the best-known Danish scientist and was evidently targeted for assassination.

Krogh spent the night in the laboratory and the next morning rode his bicycle to the home of his sister near Fredensborg, a small town north of Copenhagen. He applied for an exit permit to visit his daughter in Sweden, citing her husband's progressive kidney failure as the reason for an urgent family visit. Although such permits were controlled by the Germans, one authority was often unaware of what another was planning, and the exit visa was rapidly granted. Krogh managed to leave Denmark openly without resorting to the resistance movement's underground channels.

We had no way of knowing if Krogh had safely escaped until we received, to our great relief, a letter from him, mailed from Stockholm. He explained that the German border control had confiscated his reprints and manuscript notes, material that was returned to Copenhagen for examination by the censors. They would have a hard time, I thought, for Krogh's handwriting wasn't easy to decipher.

Before Krogh left, he asked me to take care of his personal affairs and gave me a bunch of signed checks with which to pay bills. When I wrote to him, he answered as promptly as possible, but because all letters were censored, his replies were delayed for weeks. When they finally arrived, they were marked by long colored streaks, used to test for invisible writing. The lengthy delays also ensured that any hidden information about ships, trains, freight, or German troop movements had become irrelevant.

These delays were a great nuisance because I had much to take care of for Krogh: bills to be paid, accounts to be balanced, and applications to be submitted for transfer of funds to Sweden for Krogh's living expenses and to the tax authorities for late filing of income tax returns. What normally required little time became immensely complex under a growing bureaucracy, complicated by the conditions of war and occupation. My life seemed to involve an endless series of applications, documents, and declarations.

In one of my letters I asked Krogh to contact Hevesy, who was also in Sweden, about a pay increase for one of Hevesy's young research assistants. The young man lived in two rooms on the fifth floor of a walk-up apartment building with no bathroom and only a single toilet that was shared with other apartments. He had recently married, and his wife was pregnant, but he had been too unassuming to write to Hevesy about his finances.

A couple of weeks later I received from Hevesy a long, handwritten letter telling me exactly which channels to go through and whom to contact in approaching the Carlsberg Foundation for funds. Hevesy expected difficulty only from the chairman of the board, the physical chemist Niels Bjerrum, who, he wrote, would not appreciate the importance of biological isotope research. The person to convince Bjerrum was Rehberg, although, Hevesy added, "It will not be easy." The raise came through, and I was gratified that my campaign had succeeded.

My most complex task, however, was overseeing the completion of Krogh's retirement home. Located in Gentofte, a suburb north of Copenhagen, the house would function as a combined laboratory and residence. The Insulin Foundation had supported its construction, a well-deserved reward for the man who without any personal gain had initiated insulin production in Scandinavia.

When World War II started, the new house was completed except for such inside work as doors, floor coverings, heating plant, and other installations. Work on the house stopped until 1944, when Allied forces landed in Normandy and we knew victory was inevitable. Krogh now wanted the house finished, and it became my job to oversee and coordinate the building activities, ranging from conferences with architects and engineers to discussions with carpenters, plumbers, and electricians.

There were always new questions and new problems. Could a tree be cut down? Should the walls be painted, and if so, what color? Could we plan for oil heating when there had been no oil in the country for five years? Could we substitute a composite floor for the wood parquet that the architect had planned for the dining room? Doors and lumber for inside woodwork had been acquired before the war and were brought from storage, but there was no way of finding a substitute for parquet floors, which Krogh simply hated.

More important than the many technical problems were Krogh's future plans. Without his wife's help, August Krogh needed a house-keeper for his new house. Bodil did not consider this a major problem—it shouldn't be difficult to find a competent housekeeper. Krogh therefore surprised us when, shortly after his seventieth birth-day, he wrote that the future of the house in Gentofte depended on a reply from Miss Lindberg, a laboratory technician who had worked for years as Marie Krogh's research assistant. If Lindberg's reply was neg-ative, Krogh wrote, it would be meaningless for him to complete the house. Instead he would sell most of the furnishings in his old residence and move to a small apartment in Aarhus and work with Professor Einarson in the anatomy department.

I knew Miss Lindberg well and addressed her informally as Lind-berg, following the custom among good friends. She was an intelligent woman in her thirties, always cheerful and lively, but Bodil and I didn't understand why Krogh's plans should depend so completely on whether she would be his housekeeper.

We wrote to Krogh, suggesting alternative arrangements; we would live in the new house with him until his other daughter, Agnes, re-turned to Denmark.

Krogh's response, on December 10, 1944, stated: "Your letter of De-cember 3 makes it obvious that you have not understood the severity of the blow that has hit me [presumably indicating a negative reply from Lindberg]. Well, that could hardly be expected."

Only later did I understand. Lindberg told me that Krogh had sent her a long formal letter proposing marriage and suggesting they have children together. Lindberg found it ridiculous. How could Krogh expect a positive reply from a young woman with whom he had exchanged little more than daily greetings? Moreover, Krogh was evidently unaware of Lindberg's close friendship with Dr. Rehberg. It seemed that Lindberg was the only single female Krogh knew, and, perhaps thinking his fame was enough, he had made all his plans depend on her reply. Fortunately, a month later Krogh changed his mind and resumed his interest in the new house.

No mention was ever again made of Krogh's threat to abandon the house. In fact the years he spent in Gentofte were highly productive. Until he died in 1949, he and a gifted young researcher, Torkel Weis-Fogh, who later distinguished himself as professor of zoology at the University of Cambridge, carried out pivotal studies of the migratory locust and the energy needed for their long flights. This was a question that had intrigued physiologists since the mid-1800s and had remained one of Krogh's favorite problems.

D URING THE spring of 1945 the war was nearing its end, but nobody knew how the German military would react to defeat. Would they make a last, desperate attempt at fighting, would they burn and destroy, or would they surrender peacefully? Luckily, in May 1945 the troops withdrew without armed engagement. The absence of blood-shed and destruction added to the immense relief we felt. Bodil and I went to the city, where throngs of jubilant people filled the streets. Danish flags flew everywhere, and the large square in front of the National Theater was so packed that we couldn't move.

After five years of foreign occupation, of surveillance by secret police, of insecurity and material deprivation, we were again in a free country. It was the miracle of our dreams. After the first exuberance, the return to normal conditions was slow. Even though Denmark was less ravaged than other occupied countries, a tremendous effort was required to restore the country to normalcy.

A year later I finished my doctoral work on the absorption of fat in the intestine, earning my degree in June 1946. In Denmark the final event for a doctoral degree is set in the main ceremonial hall of the university. The candidate and two faculty "opponents" wear evening dress with white tie and tails for a formal ceremony open to the public and usually attended by friends, family, and other interested persons. The morning papers report the event and usually include a drawing or caricature of the nervous candidate.

Called a disputation, the ceremony gives rise to lively discussion and vigorous criticism. My two faculty opponents, Rehberg and Hans Ussing, levied both criticism and praise, and properly so, for my work was neither distinguished nor bad. In 1948 I had the satisfaction of seeing it noted in *Annual Review of Biochemistry*, which called it a "timely" and "much-needed appraisal of the present status of the mechanism of fat digestion and absorption." The review not only discussed my work in greater detail than I felt I deserved; it also introduced my research to an international community of scholars.

Part II

The Rat That Never Drinks

[1946–1962]

Crossing the Ocean

AFTER THE WAR, it took a long time for travel between the Scandinavian countries to return to normal. Crossing borders required special permits, usually reserved for officials and other persons involved in restoration activities. Nevertheless, with the assistance of the Norwegian writer Johan Borgen, I was able to visit Norway in late May 1945, only a few weeks after the defeat of Germany. Borgen had arrived in Copenhagen as interim attaché to reopen the Norwegian embassy, and he simply appointed me a courier for the embassy, issuing a letter that served as a diplomatic passport, permitting me to cross otherwise restricted borders.

I spent only a few days in Oslo, where I discussed the possibilities for a university position. The university had barely reopened after the German occupation, and nobody could say much about the future. I returned to Denmark by train through Sweden, where to my surprise I found that chocolate was not rationed. I filled my shoulder bag with chocolates and, with the help of my diplomatic "passport," avoided both Swedish and Danish customs control.

Three months later, in August, I was able to visit my parents in Trondheim. I learned more about their difficult life during the war years. In addition to the constant problems with food, several rooms in their house had been requisitioned by the authorities to house persons assigned to live there. Having strangers in the house and sharing the only bathroom with them had been almost intolerable.

The following summer, travel conditions were more normal, and I took my family to visit my parents, who had never met Astrid and Bent. We spent several weeks with Mother and Father on the mountain farm that had been our regular stopover on the way to the lake where we spent summers fishing. Astrid was five and an affectionate little girl with curly blond hair, playful and ready to smile. Bent, at two, was quieter and more thoughtful, blond like his sister, but with straight hair.

Bent had suffered from repeated ear infections. At this time, before antibiotics, the only treatment was to keep a sick child warm and in bed. Doing so was difficult in our nearly unheated home, and Bent had been hospitalized for weeks. The hospital policy was to allow only the briefest visits by the child's mother, a regulation imposed to minimize "unnecessary" crying. Bodil went to see Bent as often as allowed, each time saddened that he smiled less and was increasingly withdrawn.

Bent had been a happy and cheerful child, curious and lively, but the long hospital stay had left him pale and passive. The visit to Norway was of great help. During the warm summer days with my parents, he regained his appetite and spent many hours walking around outside holding hands with his grandfather. I was relieved to see Bent slowly regain some strength and start to smile confidently at adults.

That summer brought other good news. I received a long telegram from Pete Scholander, a Norwegian physiologist who had spent the war years in the United States. He and Laurence Irving, an American physiologist and professor at Swarthmore College, planned to visit August Krogh in Copenhagen and then come to Norway to visit us on the farm.

This was exciting news. Both men were known for their studies of diving animals. I had met Scholander in 1938, when Krogh had invited him to Denmark. In a superb lecture he had provided simple and logical answers to questions such as "How can diving seals and whales get enough oxygen?" and "Why don't they get diver's disease, as human divers do?"

During the war Scholander had worked for the U.S. Air Force to develop simple methods to test equipment and survival gear. He tested oxygen masks for airplanes and determined the conditions under which field stoves caused carbon monoxide poisoning in tents or snow houses. He also tested sleeping bags under blizzard conditions and life rafts during winter storms in the Aleutian Islands.

Scholander and Irving were coming to Europe to discuss a proposed institute at the University of Oslo for research in nutrition, biochemistry, and animal physiology. Although I was keenly interested in the new institute, these were dreams about a distant future. The proposed institute would require careful and detailed planning, and the construction of a new building in the war-depleted country would probably take years.

When Irving and Scholander arrived at the mountain farm in their jeep, everybody found them totally charming. Irving was in his early fifties, his short, dark hair and mustache sprinkled with gray. Still youthful, he transmitted an engaging warmth and could make us all laugh with just a couple of words. Scholander was ten years younger and brimming with energy. In addition to being an outstanding physiologist, he was an internationally recognized lichen specialist. He immediately hiked up the mountainside to look around, keeping a sharp eye out for mushrooms, flowers, and lichens. I knew the flowering plants of Norway well, but not the lichens, and in a few days he taught me more about them than many botanists learn in a lifetime.

Irving offered me a position as a research associate at Swarthmore College for a year, with a similar position for Bodil. This was indeed a golden opportunity. We were excited about the prospect of traveling to America, the land that had saved the civilized world from Hitler's military hordes. The appointments would give us unmatched research opportunities with an expectation that during our time there the situation in Norway might be clarified in regard both to jobs and to housing for a new family. We immediately accepted, though we would not be able to leave for the United States before November, given my teaching responsibilities in Oslo and Bodil's research in Copenhagen.

At the end of the summer I traveled to Denmark with my family and then went back to Oslo to fulfill my teaching obligations as a university fellow, leaving Bodil and the children in Copenhagen. The housing situation in Norway was desperate, and my home during that time was a small attic room in the old Marine Biology Institute, where I slept on a hard, narrow sofa. I missed my family, and as there was no chance for them to join me, I requested permission to compress my one-semester course into a month, giving one lecture per day, six days a week. I worked hard to stay at least one day ahead of the students in my own reading. I detested being alone in the empty attic room and was overjoyed to return to Copenhagen when the lectures were over.

In October 1946, on the basis of her research on the process by which dental caries is formed, Bodil became the first woman in Denmark to be awarded the Dr. Odont. degree. Her dissertation was praised as a fine contribution to a field that was not yet well understood.

Our visit to the United States was a great adventure for which we prepared eagerly. We applied for visas and underwent the required medical examinations, chest X rays, fingerprints, and tests for venereal disease. We also had to find a good housekeeper-nursemaid because we would both work full-time. Fortunately, there were a great many girls eager for a chance to visit America. Bodil hired Inger Smith, a likable young woman in her early twenties.

One serious problem remained: all passenger liners to America were fully booked for years ahead. I contacted every shipping company in Norway and Denmark and received politely negative replies. It seemed we were stuck in Europe with no prospect of leaving when a telegram arrived from the Norwegian Directory of Shipping, advising that cabin space was reserved for us on a freighter that was unloading coal near Sarpsborg, Norway. We rushed to complete the last preparations, said goodbye to family and friends, and took the train to Norway.

On arrival in Sarpsborg we learned there was no hurry: the ship had to take on ballast and would not sail for several days. The captain allowed us on board, but then the gangway was removed and we had to remain on the ship. Noisy cranes operated all day, and the entire ship was covered with coal dust. When I had an errand ashore, one of the ship's loading cranes lowered me gently to the pier. We remained in harbor for a week, enduring dirt and grime and the narrow quarters, but savoring hot water and fruit and juices, luxuries we hadn't enjoyed for five years.

Our ship was of the Liberty class, mass produced by the United States during the war to add tonnage for transportation of troops and material to Europe. More than 2,700 were built in four years; at the peak of production, 3 ships were launched every day.

Once in transit, we ran into a heavy storm, and for twelve days the ship rolled heavily: enormous waves lifted it high up, then the ship would nosedive until its flat bottom landed with a thunderous crash before it climbed the next wave. Lying on my berth, holding on so as not to fall out, I watched our towels sway on their hooks and our shoes race back and forth across the cabin floor from one wall to the other. Astrid and Bent managed better than their seasick parents, and Inger, who was sick only the first day, took care of the children while Bodil and I suffered, our nausea exacerbated by the smell of oil from the engine room.

In the middle of the Atlantic a radio message informed the captain that a strike would prevent us from loading coal in Baltimore or Newport News and redirected the ship to Port Arthur, Texas, prolonging our voyage and entailing travel expenses I had not planned for. When the ship neared North America we turned south, the weather improved, and we were soon in the warmth of the Gulf Stream. Now the captain told us that another Liberty ship had broken apart and sunk in heavy weather in the North Atlantic. We were grateful that he hadn't told us until we were in calm, sunny weather near the Bahamas.

On December 2, 1946, twenty days after leaving Norway, we arrived in Port Arthur, greeted by warm sunshine, a delightful contrast to the wintry weather we had left behind. I stood at the railing, amazed that an industrial harbor would have such a rich bird life of gulls, great blue herons, and white egrets. I had never before seen live pelicans; they looked ungainly when sitting, but in the air they sailed gracefully on outstretched wings and dived by suddenly dropping into the water from high up.

Customs clearance proved surprisingly easy. The inspector asked,

"Do you have anything to declare?" so I explained that we were moving to the United States for a year and had only personal effects and household items. He didn't even look at our baggage.

The next person in line, a Norwegian student, expected to be admitted without difficulty. His father raised silver and blue fox, and because obtaining foreign currency after the war was nearly impossible, he had brought several beautiful blue fox pelts, which he planned to sell to pay for his studies. When the customs inspector explained that fur was subject to high import duties, the student despaired. He had virtually no money with him and no way of raising cash in Texas. The customs officer said he was compelled to confiscate the furs and reluctantly asked the young man to take the skins to the customs house. A little later the student returned with the furs, looking radiantly happy. The customs officer had searched big volumes of regulations and could find customs duties only for silver fox. As he found no mention of blue fox, he decided that these were not subject to any import duty. Once such formalities were over, we traveled by bus to Beaumont, Texas. From there we would go by night train to New Orleans, stay there for two nights, and finally take the train to Philadelphia.

In New Orleans all five of us were placed in one nearly empty room the size of a small ballroom with one real bed and several cots. By comparison with war-torn Europe, we thought postwar America was paradise. The weather was mild, and we walked the strange, palm-lined streets, looking for a restaurant. Near the waterfront I saw a sign, "Bar and Restaurant," and we all trooped in. There were no tablecloths, and the room was dark. But we thought perhaps all American restaurants were this way. A huge, tough-looking man walked over to us and in a surprisingly soft voice asked what we wanted. When I explained that we were looking for a restaurant, he said, "This is not the right place. Try the drugstore across the street." Strange to eat in a pharmacy, we thought, but nevertheless we walked in and encountered our first meal of hamburgers and milk shakes.

A day later we boarded the night train to Philadelphia, impressed by its speed, comfort, and relative quiet. In the dining car a full three-course dinner cost $1.35, including milk or coffee, and the waiters gave the children separate plates so they could share our huge portions at no charge. Although we couldn't get sleeping accommodations, there were enough empty seats for the children to lie down all night. When the train arrived in Philadelphia on the second morning, Pete Scholander was standing on the platform. From there he drove us the fifteen miles to Swarthmore, where Irving welcomed us as temporary guests in his house.

Irving lived in a rambling, two-story wooden house. One huge room occupied much of the first floor. On the second floor an inside balcony ran all around and allowed us to look down to the first floor, and from the balcony innumerable doors opened to rooms around the periphery. The addition of five newcomers appeared to make no difference in the smooth functioning of this unusual house.

After several comfortable days, we moved to our own quarters, an apartment rented from the college. Soon afterward I had a conference with Irving about my work. Irving thought an investigator should be independent of laboratory facilities so that physiological problems could be studied anywhere, and we agreed that my first project would be to develop a field method for the determination of chlorides, a method important for studies of survival at sea.

From my experience with marine birds on the coast of northern Norway, I knew the importance of simple and reliable field methods. In a few weeks I had developed a suitable method and had assembled the glassware, micropipettes, and chemicals needed for several hundred chloride analyses—all neatly packaged in a small portable kit.

The method for chloride analysis would be helpful not only in understanding problems of survival at sea, but also in other studies of lack of water and dehydration, such as survival in a desert without water. For such studies it would be useful also to have a method for determining the osmotic pressure in blood or urine.

Existing methods were imprecise or slow or required sophisticated instrumentation, so I tried to develop a better method, based on the weight change in a tiny droplet of test fluid surrounded by a known water vapor pressure. I would suspend the droplet from a fine quartz fiber so that a change in weight would be shown by the deflection of the fiber, a method that in theory should be both fast and accurate. However, it never worked well because of interference from static electricity. As so often in science, things don't always work out as expected, but I was consoled by the good results of the chloride method.

One day Irving commented, "It isn't enough that the chloride method works in the laboratory. We need to show that it works under field conditions. To test it, either we can go to the sea coast, where salt and water problems are obvious, or we can go to the desert, where there is little or no water at all."

I was familiar with the sea, having studied marine birds and osmoregulation in crabs, but I had never seen a desert, so that was where I wanted to go. This proved to be a significant choice, for it launched a series of studies of desert physiology that over the years became well known. Not until a decade later would I return to the problem of whether marine birds drink sea water; for those studies the chloride analysis was essential, but for the problems of desert animals it was less important than I had expected.

We decided to explore possibilities for working in the Arizona deserts. In March, the day before Scholander and I were to leave in the jeep, a physiologist from Oslo, Reidar Wennesland, arrived at Swarthmore. Sensing adventure, he decided to go with us. Irving, his daughter Susan, and Bodil, who was two months pregnant, flew out to Arizona, leaving Scholander, Wennesland, and me to drive the jeep.

We departed from Swarthmore on a beautiful spring day, unprepared for the blizzard we ran into in the Allegheny Mountains. We passed stalled cars and a couple of buses stranded at the wayside. A police car tried to flag us down, but Scholander put the jeep in four-wheel drive, swerved around the roadblock, and with a big smile happily waved to the cops.

The canvas top of the jeep flapped in the wind, there was no heater, and snow blew in through every crack. Wennesland and I huddled in sleeping bags as Scholander drove through the howling wind. We didn't stop until we had crossed the mountains and found a motel for the night. The next morning we woke to mild spring weather and the sun shining from a clear blue sky.

Thereafter we drove day and night, alternating at the wheel with one of us always asleep in the back. In western Texas the landscape became drier and more mountainous. When we crossed the Rio Grande, it was no more than a trickle of dirty water in the middle of an enormous dry riverbed.

Closer to Arizona the changing landscape excited me. It was increasingly dry, with barren, rocky soil and a few scattered plants. As we neared Tucson we encountered giant saguaro cactus, easily more than ten meters tall. I was unprepared for these giants and the strange desert vegetation. I had expected a barren and lifeless environment, but the Arizona deserts had a varied and interesting vegetation and a rich animal life. We arrived at our destination, having covered the 4,000 kilometers (2,500 miles) in three days.

There were challenging biological problems in this extraordinary place. How can plants tolerate periods of drought that last for months? How can any animal survive when there is no water to drink for half a year or more?

I had seen some poorly substantiated reports that kangaroo rats needed no drinking water, but this seemed impossible; other authorities claimed that these rodents depended on the moisture in foods such as cactus or underground bulbs and tubers. I was intrigued by these conflicting opinions. Could desert rodents really live on dry food with nothing to drink, and if so, by what unknown physiological mechanisms did they survive? This mystery begged for answers.

Irving supported my wish to spend the summer doing research in the desert. We obtained permission to work at an experimental station of the U.S.D.A. Forest Service located about fifty kilometers south of Tucson in the foothills of the Santa Rita Mountains. This beautiful place was ideal for our purposes, with a cabin to live in and a large

empty room we could equip as a laboratory and thus be independent of local facilities.

With these arrangements made, we headed back to Pennsylvania—Irving and Scholander by air, and Wennesland, Susan, Bodil, and me by jeep. We took roads through northern Arizona so that we could see the Grand Canyon and the Petrified Forest. The Grand Canyon left me stunned, yet peculiarly unmoved. I was somehow incapable of comprehending its magnitude. Standing on the southern rim we could see only a glimpse of the Colorado River at the bottom, the stream that has carved a mile-deep chasm in the Earth's crust, revealing two billion years of geological history.

In contrast, I found myself deeply moved by the Petrified Forest. There are no standing trees in this "forest," only huge petrified trunks lying on the barren ground. Some of these giants are more than a meter and a half in diameter and upward of thirty meters long, all solid agate in every imaginable shade. These are fossilized remains of tree trunks once carried by floods and lodged in sandy deposits. For millions of years, water seeping through later deposits of volcanic ash became enriched in silica that slowly turned the wood into hard, pure agate. Erosion then removed the softer layers of ashes and sand, leaving exposed the semiprecious wood-turned-to-stone. I was overwhelmed by the sight of this immense assembly of fossil wood, millions of years in the making.

Back at Swarthmore we spent the spring preparing for the upcoming summer in Arizona, assembling and packing equipment. The entire family then drove to Arizona, where two Swarthmore students, Adelaide Brokaw and Howard Schneiderman, joined our studies of kangaroo rats, so named because they jump on their long hind legs. The similarity to kangaroos is superficial; they are rodents and no more closely related to kangaroos than any ordinary rat. Catching the animals was easier than I had expected. We borrowed live-traps from the University of Arizona and set thirty at night, baited with rolled barley. The next morning we usually had live animals in more than half the traps. Kangaroo rats are the gentlest wild animals I have ever encountered. I could remove them from the traps barehanded and was never bitten. If I held one gently for a few moments, it quieted down and sat still on my hand.

In contrast, when our traps sometimes caught ferocious pack rats, they would bite right through thick leather gloves.

We first set about determining if the kangaroo rats stock up on water when it rains, storing a supply somewhere in their bodies to draw on during the months of dry weather. But the water content in the bodies of the animals we caught was between 65 and 66 percent—a normal amount for a mammal, indicating they were not storing excess amounts.

We then kept them in captivity, feeding them only dry rolled barley and no water. After several weeks on this diet they maintained the same unchanged water percentage. During the good life in captivity with ample food they gained weight but showed no change in their overall water percentage. An unchanged percentage of water, coupled with a weight gain, could only mean that the total amount of water in the body had increased.

That gave us the first answer: the animals remain in water balance without a drop of water to drink. But how could this be? How could they increase their total water content? We knew they were constantly losing water as they produced urine and breathed, losing water from the lungs. How were these losses covered? Of one thing we were certain: an animal cannot fabricate water out of nothing. But where was the water coming from? The only possible source was the process by which food is broken down. Known as oxidation, this process produces metabolic or oxidation water, analogous to what we see on a cold morning as moisture drips from the tailpipe of a car; water is formed by combustion of fuel in the engine.

The amount of oxidation water produced is constant, reflecting the composition of the food, be it starch, fat, or protein. For example, one gram of carbohydrates, such as starch, yields 0.6 gram of water, whether in a kangaroo rat, an ordinary rat, or a human, and nothing changes this ratio. While such a small amount is insufficient to sustain a human or an ordinary rat, it is quite enough for the kangaroo rat, which is extremely economical in its use of water.

If kangaroo rats obtain their only water from oxidation, couldn't an ordinary rat obtain enough water by simply eating more and metabolizing the food faster? The answer isn't as simple as it seems, for a greater metabolic rate requires more oxygen, which in turn requires increased ventilation of the lungs, so that water is lost through respiration; the more the animal breathes, the more water evaporates.

Ordinary air always contains some water vapor, and this affects the amount of water evaporated from the lungs. We found that in very dry air a kangaroo rat lost slightly more water from the lungs than it gained in metabolism, but if the inhaled air contained more vapor, less water was lost in breathing. When the air humidity rose above 25 percent at 25°C (77°F), evaporation from the lungs was less than the metabolic water formed, and a kangaroo rat achieved a positive water balance.

Kangaroo rats are nocturnal. They stay in their underground burrows during the day and come out only at night, when the air humidity is higher. Inside their burrows, where they spend the day, the relative humidity is higher than outside and therefore more favorable for their water balance.

We also found that kangaroo rats reduce water loss by producing highly concentrated urine. In addition, water loss in the feces is minimal. Kangaroo rats, like many other rodents, habitually practice coprophagy; that is, they re-ingest their feces directly from the anal opening. This second passage through the digestive tract greatly reduces the bulk of undigested material, thus minimizing the total amount of water contained in the very dry fecal pellets when they are finally dropped.

It took several years of careful measurements before we had all these answers, but during that first summer in Arizona we laid the groundwork for further investigations. We worked well, and Bodil, though pregnant, worked as hard as any of us. By summer's end we had a clearly laid-out plan for further research on the kangaroo rats; furthermore, the research had kindled in me a permanent interest in the many other problems encountered by desert animals. One major question that intrigued me was how camels, which have no burrows in which to escape the daytime heat, are able to manage in the hottest deserts of the world.

The summer was interesting in many other ways. We learned to pull the beds away from the wall and to keep bedding off the floor so that scorpions couldn't climb in with us. Likewise, before putting on our shoes in the morning we shook out any scorpions that had crawled in. Before taking a shower, we looked for scorpions that might have crawled up the primitive drain. Outside we watched out for rattlesnakes. Fearing for the children's safety, I killed several near our cabin. I dried the skins and cooked the firm, white meat; it had no pronounced flavor but was nonetheless exciting to taste.

Astrid and Bent liked Inger, who watched over them, conscientiously looked out for snakes and scorpions, and cooked our meals. Our laboratory was only a few steps from the house, so we could eat all meals as a family, in which we included our two students. Although we all worked hard, fieldwork also allowed us time to relax and be with the children.

The Santa Rita Mountains soared above us to an altitude of nearly 3,000 meters. Below us stretched stony and gravelly rangelands, sparsely covered with desert vegetation. In the canyon above us grew oak trees. After a late summer rain, torrents of water rushed down from the mountains and filled a dam just above the houses where we could swim and cool off.

One morning I had a high fever and was too sick to get up. A doctor in Tucson said I had pneumonia and prescribed penicillin. It made no change, except I developed an allergic rash and discontinued the drug. Over a week I gradually improved and slowly returned to work. While I was ill, Bodil worked hard and long hours. She and the two students successfully completed work that was under way; without her drive, the summer's work would have been only half finished.

We returned to Swarthmore at the beginning of the fall term, bringing with us live kangaroo rats so we could continue work on some of the unsolved questions. We had demonstrated beyond doubt that kangaroo rats can live without drinking while eating only dry seeds. Now we had to unravel the mystery and account for each component in their water balance.

During that productive year we wrote several papers describing the summer's work and the follow-up studies. Irving sent the papers to Detlev Bronk, editor of the *Journal of Cellular and Comparative Physiology*. Evidently he was too busy to read our manuscripts, for we received no reply. After several months Irving phoned Bronk, and our papers were immediately accepted without changes or comments.

Our third child was born on October 9, 1947, a sweet little girl, Mimi, blond and blue-eyed like her siblings. Intent on her research, Bodil was soon back in the laboratory, leaving Inger to take care of the children and the household.

That winter the laboratory bustled with activity. Irving was making plans for a research laboratory at Point Barrow, in northernmost Alaska, where he and Scholander would study animal adaptation to extreme cold, investigating such factors as metabolic rates, heat production, and the insulation value of fur.

Bodil and I refined our measurements of oxygen consumption and evaporation from kangaroo rats. The following summer we would return to Arizona, and in the fall we would move on to Stanford University, where Irving had arranged for laboratory space for us. With Irving and Scholander in Alaska, this would be a more stimulating environment for us than the deserted research laboratory at Swarthmore.

During our last winter at Swarthmore we were indeed fortunate to continue working with interesting problems, unhindered by lack of money and equipment. For our colleagues in Europe, the situation had improved little since the war, and when our friends asked for help with laboratory items, I tried to help as much as I could.

One request came via August Krogh from the Austrian zoologist Karl von Frisch. Von Frisch needed a polarizing light filter to find out if honeybees navigate by using the polarization of light. He had earlier shown that bees use the position of the sun for orientation, and he now suspected that even if the sun is hidden behind clouds, a visible piece of blue sky can give the bees information about the position of the sun. He would use a polarizing filter to change the plane of polarization, and if the bees shifted their flight at an angle corresponding to the change in the plane of polarization, his hypothesis would be corroborated.

Letters I wrote to the Polaroid Corporation remained unanswered. Then I remembered that a nearby auto-supply store was selling polarized windshield screens to reduce the glare from the headlights of oncoming cars at night. The screens were not only useless but in fact dangerous because they decreased rather than increased visibility. However, the screen I sent to von Frisch was exactly what he needed, so I like to think that in a minor way I helped in the striking discovery that bees can orient according to the polarization of the blue sky.

During that winter we saw a great deal of Howard Schneiderman, who was one of the brightest and most enthusiastic students at Swarthmore. He was disappointed that the college offered no biochemistry course, so I volunteered to tutor him, and during our long evening discussions I learned a great deal of biochemistry. This knowledge proved helpful when, a year later, I started lecturing in Oslo, where I was appointed a lecturer in biochemistry and physiology.

At the beginning of the summer of 1948 we returned to Arizona, excited at the prospect of adding another piece to the puzzle. During the winter we had established that the water balance of kangaroo rats depends critically on the air humidity. We now wanted to study the humidity inside the burrows where the animals spend their day and find out whether our laboratory results were compatible with real-life conditions.

We had a small temperature and humidity recorder, the size of a wristwatch, which recorded air temperature and humidity continuously for twenty-four hours. To get the recorder to an undisturbed nest chamber, we enlisted the help of the animals. We trapped a kangaroo rat during the night, and in the morning we tied the recorder to its tail and let it loose. It headed right for home, pulling the recorder to the nest chamber. A day later we excavated the burrow, retrieved the recorder, and released the animal. The results established that the humidity in the burrows is substantially higher than in the outside desert air, a factor that our laboratory measurements had indicated was essential for the rats' survival.

After a fruitful summer we continued to California, where we settled in a comfortable new house in Palo Alto, near the Stanford campus. Irving had arranged for us to work for a year as guest investigators in the Physiology Department of Stanford Medical School while retaining our postdoctoral positions at Swarthmore.

I had also accepted a position at the University of Oslo, and in early September I left for Norway to take up my teaching duties. The university permitted me to compress a semester's teaching into about two months of lectures and then return to Stanford for the remainder of the year to continue my research. I was to teach biochemistry and physiology concurrently, and because there were no teaching laboratories for these subjects, I concentrated on giving good lectures. Each day, six times a week, I lectured in each subject, much as I had two years earlier, trying to keep one step ahead of the students.

The housing situation in Oslo remained desperate, and I again lived in an attic room in the old Marine Biological Laboratory. Shortly before returning to California in late October, I had a conference to discuss housing prospects with the university secretary, the top administrative officer of the university—a fat, pale man sitting at an enormous polished mahogany desk. He explained that under certain conditions the university could bypass the long waiting lists for housing. It turned out that the director of the university zoological museum was retiring and moving to his country farm, thus freeing his former residence. By shuffling people around, an apartment would be made available for my family in an older part of the city.

This was welcome news, until the secretary with an obliging smile added: "I hope you do not object to facilities across the backyard." For several seconds I didn't understand what he meant. Then the meaning hit me: the apartment had no indoor toilet! I had no idea that Oslo still had apartment houses that lacked modern plumbing. It was almost unbelievable, but people in Norway after the war were happy for any sheltered place. I thought of our comfortable house in California, where my children played in the sunshine, and of the long, dark winters in Oslo, colds and ear infections, and no indoor plumbing.

I walked out into the gray and rainy day, my enthusiasm chilled. I scrapped any immediate plans for moving my family to Norway and decided to resign my position. I finished my lectures for the term and flew back to California.

During our year in California we felt welcome and respected in the Stanford Physiology Department. Publications coauthored by Bodil and me began to appear, causing somewhat of a sensation, for nobody thought that a mammal could live on dry food without any water to drink. We reported our findings at scientific meetings and became recognized as a team.

We knew exactly what measurements we needed in the continued study. We solved some difficult technical problems and accumulated all the information required to give a complete account of the water balance of the rat that never drinks.

One major test was to keep kangaroo rats at different air humidities to see if our predictions about water balance were right. Instead of buying expensive equipment, I put together some simple components to secure air at known, constant humidities. The results validated our calculations, and I summarized our findings in an article published in 1951 titled "A Complete Account of the Water Metabolism in Kangaroo Rats and an Experimental Verification." It thrilled us that Homer W. Smith, the eminent renal physiologist at New York University, remarked that our studies represented the most complete single investigation ever conducted of the water metabolism of any animal.

We still could not explain why evaporation from kangaroo rats was much lower than that of white rats or mice. We had no doubts about the accuracy of our measurements, but we could not explain the mechanism. One possibility was that the oxygen extraction in the lungs was greater than in other animals, which would reduce evaporation relative to the amount of oxygen used. However, this should be reflected in a high oxygen affinity of the blood and in its acid-base balance. This hypothesis, it turned out, was wrong, and not until years later did I have the explanation.

A Norwegian biochemistry student, Björn Gjönnes, helped me in this work. In reviewing the literature on oxygen binding in blood we found a previously unnoticed relationship, that small animals generally have a lower oxygen affinity of the blood than larger animals. This phenomenon could not readily be explained and led to my long-lasting interest in how structures and physiological processes are scaled in relation to body size.

The children thrived in the sunny California climate, playing in their small plastic pool on the lawn behind the house. However, the eternal sunny weather got on the nerves of my young Norwegian friend. After a couple of months of unchanging weather, Björn started complaining. Every morning he came to work with the same remark: "There is no weather here!" I often wondered how he felt about California weather after he returned to the long, dark winters in Norway.

I had given up all hope of moving to Oslo within the foreseeable future, but staying at Stanford was out of the question because Irving had been unable to obtain continued funding for our research. Bodil didn't want to remain in the United States, but I couldn't see returning to Scandinavia, where we didn't have jobs, while in the United States the chances for finding grant support for research seemed excellent. Shortly thereafter I saw in the journal *Science* an advertisement for an investigator with experience in fat chemistry. Thinking I had ideal qualifications for the job, I was not surprised to be offered an assistant professorship in the Department of Preventive Medicine and Industrial Health at the University of Cincinnati College of Medicine. We now had a place to move to when we left Stanford.

Learning that her father had liver cancer, Bodil left for Denmark with the children that summer to be with August Krogh during his terminal illness. I remained in Palo Alto to find a buyer for our house, pack our belongings and our laboratory equipment, and move to Cincinnati.

Personal Struggles

I HAVE NO CLEAR recollec-
tion of my last weeks in Palo Alto except that, when I sold the house
and had to move out, two biology students offered me a place to live for
a few days before I started on the long drive to Cincinnati. We had a
good time together, and I felt much like a student again, eating primitive meals, drinking California wine, listening to records, and reading
books. I plowed through two of Margaret Mead's books on native life in
the Pacific, searching for passages pertaining to the sexual mores and
customs of the islanders. In the late 1940s this was the only kind of literature anyone could legally acquire that had any direct mention of sex.

I packed most of our belongings in a trailer and then drove north
along the coast of California and Oregon to Seattle, where I visited Art
Martin, a well-known animal physiologist and professor of zoology at
the University of Washington. Turning east, I continued on a northerly

97

route to Cincinnati, stopping at Yellowstone to admire the hundreds of hot springs, then through Idaho and South Dakota, where the most memorable sight was Mount Rushmore. Never had I seen anything in such bad taste—giant portraits of four U.S. presidents blasted and chiseled out of a mountainside.

In Cincinnati I rented a room while waiting for my family to return. My clean and orderly room was only a couple of blocks from the laboratory, but I hated living alone, eating my meals in the hospital or at a small nearby restaurant.

I hadn't heard much from Bodil since she had left for Denmark, but I knew she was busy caring for her father, who eventually died, on September 13, 1949. Expecting my family soon thereafter, I was surprised when in early October I received a letter from Bodil explaining that she had been in an automobile accident. Her brother, Erik, had lost control of his car on a curve, and they ended up in the ditch. Bodil had a concussion and was to remain in the hospital for several weeks. The worry about Bodil's accident, her vagueness about her injuries, and the delay in her return disturbed me. I didn't know how seriously she was hurt, and it wasn't until months later that I learned she had no injuries other than a slight concussion and was hospitalized only as a precaution.

As the weeks went by I heard little from Bodil and couldn't understand why she had not returned to the United States upon her release from the hospital. In November she wrote that she might remain in Denmark for the winter to resolve problems with her father's estate; she also felt that it would be good for the children to stay there. Calling overseas wasn't thought of in those days, so I had to rely on her letters. In the end, to my great relief, she made up her mind and arrived with the children in Cincinnati a few weeks before Christmas.

I was looking forward to seeing them after my many months alone, and in anticipation of their return I had rented a house in a good neighborhood. I met them at the airport, and we drove to the house that would be our home for the year. When Bodil stepped through the front door, her first words were "What a terrible place this is."

"I did the best I could, and it is only temporary," I answered. I couldn't understand her hostility, or why the house wouldn't be good enough until we found another place.

It quickly became apparent that my marriage was unraveling, a situation that adversely affected my work. I had set up a laboratory for microchemical studies, but my personal problems left me without enthusiasm for my work. The scientific questions I worked on seemed trivial, and I produced only a couple of mediocre publications.

In contrast, Bodil continued our studies of kangaroo rats with great energy. The first summer in Arizona we had found that kangaroo rats have extraordinarily powerful kidneys, capable of producing urine more concentrated than known for any other mammal. I had obtained a grant to pay Bodil's salary and cover laboratory expenses. She worked hard and made good progress, which underscored the lackluster of my own activities. Although I had initiated and organized the research on kangaroo rats, she turned from a major contributor into sole investigator, no longer sharing the findings with me. In turn I became increasingly resentful, comparing her behavior with my own willingness to share with her our earlier work and all my best ideas.

Bodil often complained that because she was a woman, she was more likely to be viewed as a technician than as the full-fledged scientist she was. To help allay her concerns, I had recognized her as a major contributor by putting her name first on many of our joint publications, probably giving her more credit for the desert work than was warranted, as I had done much earlier for our spinach studies.

One day I felt I could no longer cope with my disintegrating marriage and sought the help of a psychiatrist, Dr. Fabing, recommended by a friend. After I had poured out my problems, Fabing said he wanted to see Bodil, who undoubtedly was as unhappy as I was.

We didn't have regular appointments with him, but one or the other of us called him when the situation at home seemed especially explosive. It was a relief for me to talk with Dr. Fabing, but I felt he hadn't understood our situation when he commented, "I think there is a great deal of competition between the two of you."

"Not at all," I replied, shaking my head. "We always share our research generously and without any thought of competition."

The idea had never occurred to me, but as I thought about it over the next few days, I began to see that Fabing was right. Our superficially friendly collaboration concealed an underlying ambition to excel;

each of us wanted to be the best. Recognizing our rivalry, I mentioned this to Bodil, repeating Fabing's words: "I think there is a great deal of competition between the two of us."

Bodil's instantaneous response was like an echo of my own: "Not at all; we have always collaborated so well," she declared firmly. Nothing more was ever said on the subject, and, sad to say, I don't know whether she ever understood her own part in the conflict.

Gradually I saw how my own marriage was influenced by the relationship between my parents, remembering how Mother had so often told me about working hard for little recognition and about Father's being domineering. She believed her help in his research had never been sufficiently credited. In my marriage I had tried to avoid doing similar injustice to my intelligent and hard-working wife by going out of my way to share my best ideas with her and by giving her more than full credit for her contributions. And Bodil had undeniably grown up amid expectations that she would excel. Her parents, like mine, were both professionals, and they too had collaborated on several projects in the early years of their marriage. Moreover, Bodil had been devoted to her father, who had approved of his youngest daughter's success. Living up to her father's expectations was a natural wish that she carried with her always.

The conflicts in our marriage were not limited to professional rivalry. Although we didn't argue over money or how to bring up children, there was a constant tension in all our relations. Nevertheless, despite other deep-seated disagreements, we were in many ways dependent on each other. Each of us wished to escape from the daily misery, but we managed to get along without an actual breakup of our marriage. Still, our discord definitely affected my work, and I felt I had disappointed Dr. Kehoe, the director of the institute. Although he remained cordial and generous, I certainly contributed less to the research program than he had expected.

Most of my daily contacts were with Ray Suskind, a dermatologist with whom I was supposed to collaborate on studies of skin lipids. We did publish a paper together in a dermatological journal, but I felt that nothing of importance came out of my work. I came to admire Suskind, a well-informed and fine physician, and I regretted that I was of so little use to him.

My only official teaching obligation was a brief biochemistry course for postgraduate students in industrial medicine. These young physicians had forgotten most of their biochemistry and physiology and knew little about important metabolic pathways, such as the enzymatic steps in the Krebs cycle and the production of the universal energy currency of living organisms, adenosine triphosphate, or ATP. Yet they were intelligent and highly motivated, and I found it greatly rewarding to teach them enzyme kinetics and biochemical pathways, knowledge of great importance to their work with pollutants and toxic compounds.

During the three years in Cincinnati I tried to find grant support for more productive projects than those in which I was halfheartedly engaged. My first application was to the newly formed National Science Foundation (NSF) for two projects.

One of these pertained to the relationship between the metabolic rate and body size of animals. My idea was to use the large cecropia silk moth. Intervention in the endocrine system of a larva can induce it to pupate long before it has grown to normal size, thus producing a midget adult. Other interventions can prevent the normal pupation of a fully grown larva, inducing it to continue growing before pupation and in the end producing a giant adult. These interventions, I suggested, would make it possible to study essentially identical animals, all adults, but of widely different sizes.

I also wanted to clarify whether the limit to the size of insects is set by their mode of respiration. Insects obtain oxygen through a system of fine tubes, the tracheae, through which oxygen diffuses to the tissues. It had long been thought that the tracheae set an upper limit on insect size, and I planned a series of experiments to find out if this was true. Carroll Williams at Harvard University, who had carried out spectacular research on the endocrinology of silk moths, offered to help me with the difficult endocrine techniques needed to produce giant silk moths. Although the NSF acknowledged the receipt of my application, I never received a formal reply.

I made another application for a study of the possible advantage of animal fur in hot climates. Studies by E. F. Adolph at the University of Rochester had shown that humans, when lightly clothed in a hot desert, need less water than when they are naked and the sun beats down directly on the skin. What would the situation be for heavily insulated animals, such as camels? I proposed to study sheep and determine their rates of water loss from panting and sweating when they were exposed to heat, both before and after shearing. This application met the same fate as the silkworm project.

A third application, in this case to the Office of Naval Research, evolved from my work with Ed Pelta, a young electronics student in my laboratory. We developed a method to count red blood cells automatically. Up to that time the routine procedure for counting red blood cells was to dilute a blood sample, place a droplet of the suspension in a counting chamber, and use a microscope to count visually the number of cells in each square of a grid. We proposed to let a dilute suspension of red cells flow through a capillary tube where a fine beam of light would be interrupted by each red cell, yielding an electronic count of the cells. The Office of Naval Research had expressed an interest in automatic counting of red cells, perhaps related to concern about radiation damage to the red bone marrow from human exposure to nuclear explosions. Our proposal had a reasonable probability of success, but it was turned down. Not much later the Coulter counter was developed, and it rapidly became the standard instrument for counting red cells.

During this relatively unproductive period, I tried to distract myself from my failures at work. Frank Dutra, the pathologist at the institute, became a close friend. He has always been a lively and enthusiastic person, curious about animals and plants and what goes on around him, a true biologist. His wife, Joyce, had an M.D. degree but was also an excellent artist, her paintings ranging from expressive character studies to serene still lifes. Their children matched ours in age, and we often spent Sundays together until Frank accepted a position in California. I was sorry to see him leave, for I felt closer to him than to anyone else in the area.

The constant tensions at home led me to register for a class in ceramics and pottery at the School of Engineering. As soon as I mentioned to Bodil that I was taking the course, she wanted to take it too. For a change I didn't defer. Why does she always want to compete with me, I thought. I told her to register if she wanted, but I was not going to be in the same ceramics course. There was no more discussion.

I had long been interested in pottery, I liked to work with my hands, and the evening course took me away from the stresses at home. My classmates ranged from a retired engineer to a couple of youngsters who took the course just to pass time. The instructor, Harold Nash, was an excellent teacher, and he liked the fact that I had enough knowledge of chemistry to understand what I was doing. I quickly completed assignments and started experimenting with new techniques.

One day out of the blue I received a call from a Mrs. Lazarus, who explained that the Cincinnati Symphony Orchestra had scheduled a performance of the *Gurre Lieder,* by Arnold Schönberg. The lengthy text was available only in German, she said; could I possibly help by translating it to English? The only reason I could see for asking a scientist with Danish connections, and not a professor in the German Department, was that the Gurre legend is set at the Gurre Lake, in Denmark. Whatever the reason, I said I would be glad to help.

The job wasn't as simple as I thought. A good translation should retain the style of the original while recreating the emotional meaning in English. I felt that only a genius could manage this task with the old-fashioned German text, and I ended up doing a literal, word-for-word translation that was neither good poetry nor good English, so I refused to have my name printed in the program. Perhaps this was unnecessary modesty, for mine is probably still the only existing translation of the *Gurre Lieder.*

Whatever I may have felt about the translation, the Symphony Society was pleased, and I received more requests for help. I did a better job translating Richard Strauss' *Four Last Songs (Vier Letzte Lieder).* These songs are for soprano and orchestra and are less antiquated than the *Gurre Lieder.* They touched me emotionally; the last song alludes to the inner thoughts of a husband and wife walking hand in hand so as not to lose their way. How I wished I could confidently hold the hand of my own wife.

In contrast to our miserable home life, Bodil and I were increasingly recognized for our research contributions. In 1951 the editors of *Physiological Reviews* invited both of us to write a review paper on the physiology of desert animals. Such invitations were usually extended to more senior scientists, and we happily accepted the honor.

That summer we traveled with the entire family to Woods Hole Marine Biological Laboratory to take advantage of its excellent library. We lived in a rented beach house, and as usual a young Danish woman served as our housekeeper, leaving us free to work full-time.

After long hours in the library, we had accumulated extensive notes, and I began to plan an outline for the review. As I started to write an introduction, I knew that I would be the person to write this review. The background material was well organized, and I wrote continuously day after day, fully aware of the feelings of resentment and anger that drove me. I asked Bodil to write some small pieces where additional information was needed. Then, after completing the review, I placed my name as first author on the published paper.

In working on this review, I realized that virtually nothing was known about the physiology of camels, the most celebrated of all desert animals. I could infer what physiological mechanisms would be of importance, what questions needed to be asked, and how those questions could be approached. At that time Bodil was more interested in a comparative study of various small rodents that live in deserts around the world, but my view was that other desert rodents, though not closely related to kangaroo rats, were bound to use similar physiological mechanisms to survive. The real breakthrough in our understanding of desert animals would come, I maintained, by studying camels, which undoubtedly had evolved totally different solutions to desert survival and thus might reveal interesting new physiological principles. My arguments prevailed, and we began talking about future opportunities for camel studies.

We returned to Cincinnati and our usual routine: breakfast with the children before sending them to school; then off to the laboratory, where Bodil continued her studies of kidney function and I tried, with less enthusiasm, to do my work; then home for dinner with the children, followed by reading or telling bedtime stories.

In THE SPRING of 1952 I received a letter that completely changed my life and that of the family. The chairman of the Zoology Department at Duke University wrote to ask if I would be interested in a position in his department. They needed a physiologist because one of the zoology professors, F. G. Hall, had become chairman of the Physiology

Department in the Duke Medical School and another physiologist, Karl Wilbur, was scheduled for a year's leave.

On the flight to North Carolina for an interview, I sat next to a black teacher from Cincinnati who was going to Durham to visit her family. At a stopover in Charlotte, we walked together into the small terminal building, and at the refreshment counter I ordered two Coca-Colas. The girl behind the counter looked stunned. She put two filled paper cups in front of us, took my money, and forgot to give me change. Only much later did I understand that it was unheard-of in the South for a white man and a black woman to be seen together in public, not to mention drinking Cokes. A few years later, when racial demonstrations were common, the girl at the counter would probably have refused us service.

The chairman of the Zoology Department, I. E. Gray, met me at the airport in Durham and drove me through pouring rain to an aged downtown hotel. The weather didn't improve, but the campus, with its gothic architecture, impressed me.

I lectured on the desert rats and went through the usual interview routines. It was agreed that if I accepted a position, Bodil would be appointed a research associate and would continue to work on research grants. During my final interview in the office of the vice president, Dr. Paul Gross, I made two requests. One was to be allowed a leave of absence without pay in my second year at Duke to study camels. This leave would, of course, depend on my success in raising funds to support an expedition to the Sahara Desert. My second condition was to be given the rank of full professor, an academic status I considered necessary to secure funds and to help negotiate with French authorities in North Africa. The outcome was what I wanted—the offer of a position as professor of physiology in the Department of Zoology. The title pleased me because I am primarily a physiologist and not a zoologist, although I had not emphasized this in my interviews. I was delighted to accept.

In the spring of 1952, Bodil received an invitation from Homer Smith to join for the summer a research group at the Mount Desert Island Biological Laboratory, in Maine. She took it for granted that she should accept this opportunity, and in early summer she took off for Maine with the children. Thus I was once again left with the responsibility of packing laboratory equipment, household goods, and furniture; selling our house; and moving to Durham, where everything would be stored until we returned from a visit to Europe.

In early August I joined the family in Maine, and we all traveled to Denmark, where the children stayed in the Krogh country house and had a marvelous time with their cousins. I spent much of my time in Copenhagen, where Krogh's laboratory had remained active as Torkel Weis-Fogh continued the research on locust flight that he and Krogh had started.

Torkel's wife, Hanne, was a highly intelligent, attractive woman who moved and spoke with a quiet dignity. I felt relaxed in her company, and we got along well. Talking with her was comfortable and very different from the tension I felt when I was with Bodil. Although I didn't see Hanne very often, we came to know each other well and became good friends.

Visiting my parents in Trondheim, I found that my father had become an old man. He had lost much of his vigor and strength and was severely affected by Parkinson's disease. It touched me deeply that he was weak and needed me, that I was young and free and could plan my future, whereas he had none. He was eager to tell me about the many unusual objects and pieces of furniture in the house, from what branch of the family they had come, or how he had acquired them. He was seeking to establish a continuity between the generations that he knew would otherwise be lost.

Mother also had aged, but she had retained her unquenchable energy despite the wearing effects of the war and the heavy burden of caring for Father. Although she had not seen me for two years, she didn't complain when I left or beg me to stay longer. Characteristically, she never asked favors of her grown children.

Bodil and I had been invited to participate in an international conference on the biology of deserts in London in late September, and soon after I returned to Denmark we went to England. International conferences were not as common then as they are now, and the meeting was an exceptional opportunity to become acquainted with other investigators. We listened eagerly to information on the natural history of desert animals and plants. However, nobody seemed to know much about the physiology of desert animals; we were unique in having solid answers to questions about how desert rats manage without drinking water. Bodil summarized our work on the kangaroo rats, and I outlined major problems for future research. I emphasized the problems arising from differences in animal body size, using the kangaroo rat and the camel as examples. We already knew how small desert rodents survive without drinking, but not what makes camels well adapted to desert life. I fully intended to get a closer look at these enigmatic animals.

After the conference in London I took up my faculty position and teaching at Duke. Bodil became a research associate with no teaching obligations and continued studying renal function, supported by research grants. Our relationship continued to be tolerable, though strained with ups and downs. I don't think that either of us at that time seriously considered separation or divorce; we just continued our lives unaltered. Mimi was five and old enough for kindergarten, and Bent and Astrid, aged eight and eleven, started in the Durham public schools.

Our Danish household helper, Kirsten Friis, had become a close family friend and moved with us from Cincinnati. Blond and curly-haired, she could easily have been taken for an older sister of our Scandinavian-looking children. With the children in school, Kirsten enrolled as a student at Duke, and we all pitched in to keep the household going.

As we settled down to the routine of the academic year, I spent every minute I could spare planning for a year's camel research. It was essential to select a suitable place among several areas where camels were common. An Algerian zoologist, Professor Francis Bernard, whom I had met at the conference in London, had given me much useful information. He suggested that I should contact Frode Eckardt, a Danish botanist who was familiar with the Sahara. The name, even when pronounced in French, sounded familiar; we had been fellow students in Copenhagen.

When I wrote Eckardt, he replied with a long and detailed description of places I could consider. The oasis Béni Abbès, 1,300 kilometers south of Algiers by road, sounded promising. There was a small French research station there. The oasis had a good water supply but no electricity except for a small generator in the physics laboratory. There was also a modest French military outpost with a resident physician on the staff.

To mount an expedition to the Sahara Desert would require significant financial support. I had a small grant of $1,500 from UNESCO, which represented only a tiny fraction of what was needed. Even so, this token sponsorship was of great value, for instead of speaking about the problems of raising funds for an expedition, I told friends and colleagues that, backed by UNESCO, we were proceeding with our camel research.

In addition, Bodil and I each received fellowships of $4,000 from the Guggenheim Foundation, which made it possible to manage a year's leave without pay, although my award was less than half of my academic salary. The fellowships would cover travel and living expenses for the entire family, but we needed at least another $10,000 for equipment, chemicals, a field vehicle, camels, local help, and other miscellaneous items. Funds were also needed to pay salaries and travel expenses for one or preferably two scientific collaborators to make maximum use of our time in the desert.

I applied to the Rockefeller Foundation, the Markle Foundation, the Carnegie Institution, Lilly Pharmaceutical, and others. Some replies were sympathetic, others encouraging, but none promised funds. I hoped for support from the King Ranch, widely known for developing the heat-tolerant Santa Gertrudis cattle. The owner, Mr. Kleberg, was extremely interested in the effects on cattle of heat and lack of water, but financial aid for work on camels was beyond what the ranch could support. Similar replies came from oil companies because prospecting parties no longer used camels. Our plans, though interesting, could not be funded.

Most disappointing of all was yet another brush-off from the National Science Foundation. Louis Levin, the program director for regulatory biology, seemed openly unfriendly when I went to Washington to explain our plans. "Of course you can apply for funds if you insist," he said, "but why go to Africa? What is interesting about camels anyway?" My application was turned down. A few years later, when the success of our camel studies was widely recognized, I met Levin again, and on that occasion I thought he seemed embarrassed.

With the NSF letter of rejection, we were getting desperately short of time. We planned to leave in early September and still had no funds for supplies and equipment. My saving angel was Dr. D. H. K. Lee, professor of physiological climatology at Johns Hopkins University and former professor of physiology at the University of Queensland, in Australia, where he had studied the responses of domestic animals to high temperatures.

Dr. Lee sent me a book manuscript, "Methods for the Study of the Heat Tolerance of Domestic Animals in the Field," and asked me to comment on it. I was keenly interested and phoned to ask if I could visit him and talk about the manuscript. He met me in downtown Washington, and we drove to Beltsville, Maryland, where he worked at the research facilities of the U.S. Department of Agriculture. We spent the day eagerly discussing problems of mutual interest, and he was enthusiastic about my camel plans and  wanted to help me. He was a consultant to the Office of the Surgeon General and the Office of the Quartermaster General, where he introduced me to the civilian research personnel and explained how important my plans were. They listened thoughtfully, then acted with surprising speed. Less than two months later I was awarded $8,736, sufficient for most supplies and equipment and for a year's salary of $5,000 for a collaborator. Three days later this award was increased to $15,456, allowing me to hire a second scientific associate.

The first scientist to join us was Richard Houpt, a veterinarian from the University of Pennsylvania, whose experience with animals would be valuable. The other collaborator, Stig Jarnum, was a young Danish physician interested in physiological research. It was also reassuring to have a physician on our team in case of emergencies.

The Office of the Quartermaster General supplied an old four-wheel-drive army ambulance, which I converted for our use by removing all inside equipment and covering the olive-drab exterior with aluminum paint. We also received army blankets and cots, cameras and film, antibiotics, and other medicines that would be needed if anybody got seriously ill. An expedition to a remote place requires meticulous planning; there is no way to make up for even the smallest missing detail. We had to be independent in every respect, including the generation of electricity for our instruments.

One of my big challenges was to devise a method for weighing camels accurately, for weighing is the simplest means of determining how much water is evaporated to cool the body in extreme heat. This method is used for humans; the amount of sweat evaporated can be measured by weight loss, provided the scale is accurate. Cattle and horses are usually weighed on large, heavy platform scales, which would be totally impractical for us. We needed a precision scale, not too bulky and heavy, that could measure up to 500 kilograms, and, given the desert heat we expected, it had to be temperature compensated. I decided I could use a spring-loaded scale and hoist a camel up in a canvas sling.

At an industrial fair in Philadelphia I located not only a supplier willing to manufacture the scale but also a light and convenient hoist for lifting the animal. For the sling I bought strong canvas, grommets and grommet tools for making reinforced holes, and suitable rope for threading through them. Every measurement we wanted to make required similar meticulous planning.

As with the kangaroo rats, a determination of total body-water content would tell us whether camels store extra water in their bodies, which could explain their legendary ability to go without drinking. By the same method we could tell how rapidly the body water is used when water for drinking is unavailable.

By checking blood volume we would know whether the blood becomes more concentrated as a camel becomes dehydrated for lack of water. Chemical analysis would reveal the concentrations of salts and other components in the blood. Our work with kangaroo rats had shown the importance of kidney function in water balance, so we planned to measure creatinine and inulin, standard indicators of kidney function. All these measurements required complicated methods for determining sodium, potassium, chloride, and urea concentrations in blood plasma and urine.

We made lists of all required chemicals, glassware, pipettes, centrifuge, and balance for weighing chemicals. We estimated the amounts of chemicals we needed, allowing for some to spare but trying to avoid unduly increasing the weight of the expedition. Then, when everything was on hand and checked against the lists, it had to be carefully packed to prevent breakage during the Atlantic crossing and subsequent transportation on rough desert roads.

In addition, I had to navigate the red tape that was involved in taking scientific equipment out of the country. I needed an export license, and for this I had to list each item and give its value, weight, and country of manufacture. I skimped in compiling these lists, confident that no customs official would inspect every item, listen to my explanation of its function, check its value and origin, and finally tally all items against my lists.

To ease the task of getting equipment into Algeria, I contacted the State Department for introductions to the French embassy in Washington and the U.S. consulate in Algiers. I traveled to Washington for endless discussions with officials who were neither interested nor very helpful.

The summer before our departure was hectic with preparations. I worked long hours without air conditioning in the laboratory or at home; even so, it was more tolerable than the preceding sultry summer in Cincinnati. Bent and Astrid, now nine and almost twelve, did most of the cooking and baking and took their new duties as a challenge.

By late summer everything was in order. I had booked passage on a French freighter destined for Algiers, scheduled to leave New York on September 1. In late August the agent called to say there would be a delay until September 23.

A Year in the Sahara

In LATE SEPTEMBER we were finally on our way. Bodil, Astrid, and Mimi went by train to New York while I headed north with Bent, driving the truck full of scientific supplies and our personal effects.

In New York I cleared the last formalities with customs officials and the shipping company. The departure of the ship was again postponed, but the company was hoping to be ready before a threatened strike closed down the New York waterfront. I was uneasy at the prospect of sitting in New York for an indefinite time, but we were lucky: we sailed on September 30, 1953, just before the strike broke out.

Dick and Doris Houpt traveled with us on the freighter, which carried mixed goods and a handful of passengers to Portugal and North Africa. The ship sailed almost due east on a latitude of 40° North, through the Azores, headed for Lisbon. One morning I found a flying fish that had landed on the deck during the windy night. The children were fascinated by its large pectoral fins, almost like the wings of a sparrow, which enable the fish to soar as it skims across the surface of the sea.

We arrived in Lisbon on October 9, Mimi's sixth birthday. The cook, Monsieur Cremadés, had grown especially fond of her and had baked a cake. We placed six candles on the cake, put a few small gifts on a tray, and carried it to her as a surprise in the morning. In her soft and quiet way she expressed how marvelous it was to have a real birthday, especially on a ship.

After calling at Casablanca and Oran, the ship arrived in Algiers on October 17. We moved to a hotel, and a couple of days later Bodil, the children, and Doris Houpt left by train for Colomb Béchar, where they would be met by people who would drive them to Béni Abbès.

Dick and I remained in Algiers to get the truck and our equipment cleared through customs. For the next week I explained over and over to various authorities that I was the leader of a scientific expedition, that I had the necessary documents, including recommendations from UNESCO and the French consul in Washington, that I was on a temporary visit, and that my goods would leave the country again. It was difficult to explain this convincingly in French, but my documents were always carefully scrutinized before I was sent to the next official, who supposedly would be the person to give me the permission to enter.

The problem was finally solved with the aid of Frode Eckardt, the Danish botanist who had recommended Béni Abbès to me. He was now living in Algiers and spoke fluent French, and he and I visited every existing government office until we got the equipment released. No one really wanted to prevent my expedition from entering the country, but we could not proceed until every civil servant who might be involved had formally declared himself not responsible. After a week of shuttling between countless government offices, we finally found one person who said yes. It was Saturday, and we hurried to the customs house, expecting another obstacle; but all the customs inspector did was lift the lid off one of my many boxes and let me pass. We got everything out just before the customs house closed at noon.

We were finally ready for the trip across the Atlas Mountains and south to Béni Abbès. We said goodbye to Frode and drove to Mascara, where we slept at a peculiar old hotel, a primitive one-story structure with a single electric bulb hanging from the ceiling. At dinner we carefully avoided the salad, picked at a very smelly fish, ate a little meat, and drank sour red wine.

The next day we left before daybreak and drove past Sidi bel Abbès, the headquarters of the Foreign Legion, eager to reach Béni Abbès and full of excitement about our research. At noon in Saïda, we went to the market and bought wonderful oranges, mandarins, and small, aromatic pears, then continued south.

Soon afterward, on a long uphill slope, the motor began to lose power. Dick thought the fuel supply was insufficient and cleaned the carburetor. The truck started easily and ran well, but over the next thirty kilometers it gradually got worse. Dick found the fuel pump full of gas, concluded that nothing was wrong with the fuel supply, and again checked and adjusted the ignition and timing. The engine started readily, but on the road we again soon lost power. We did not dare continue into the desert, so we drove back to Saïda and woke a mechanic who was enjoying a Sunday off. For several hours he went over everything Dick had checked—with exactly the same result. We found a hotel for the night.

The next morning the hotel owner gave me the names of the best mechanics in town. We went on foot to all of them to look them over, but none inspired confidence. Then we passed a shop where several trucks with local license plates were parked; the place seemed busier than the others. Back at the hotel the owner said that the mechanic there was not to be recommended. But we reasoned that if a mechanic was good enough for local truck owners, he could probably help us, so we drove there. It was a good choice. He meticulously checked ignition, carburetor, fuel pump, and so on; everything was in perfect condition. Then he discovered that the line from the gas tank to the fuel pump was clogged. The gasoline had made the old rubber start swelling until the lumen of the hose was reduced to pinhole size. We hadn't suspected this because, as soon as we stopped the engine, the vacuum in the fuel pump had slowly pulled in gas, so that when we opened the pump it was full of fuel, indicating that the fuel supply was in good order.

Happy that the truck was now roadworthy, we continued to Krejder, the last little group of houses before the true desert began. There we occupied both rooms of a tiny hotel. The next morning the road became little more than unclear tracks that were hard to follow. We maneuvered around and over rocks, constantly in danger of broken springs. On open stretches the tracks were like a washboard and in places obliterated by dirt and sand. To avoid the washboard surface or getting stuck, drivers had swerved around the worst places, seeking better conditions and making new tracks that spread out fanlike. Over the years the road had become a wide and ill-defined band of criss-crossing tracks, and unless one knew the landmarks, one might wonder where the road was.

Five days after leaving Algiers we arrived at Béni Abbès just before dark, tired but happy. We found the others temporarily installed in the primitive hotel, and now we occupied all its rooms.

It was my first meeting with Stig Jarnum, the Danish physician we would work and live with for the year. Boyish-looking, tall and slim, with straw-colored hair bleached by the sun, he proved to be a competent scientist, serious, always helpful, and a wonderful collaborator. His wife and two-year-old son had remained in Denmark, where the boy was recovering from whooping cough; they were to arrive a few weeks later.

Helga Neal, Bodil's cousin, arrived from her home in London, having agreed some months earlier to serve as Bent and Astrid's teacher for the year.

BÉNI ABBÈS is perhaps the most beautiful oasis in North Africa, with lush groves of date palms and small, well-kept gardens. It is located along a *wadi*—a wide, flat riverbed, dry except in winter, when heavy rains in the Atlas Mountains may bring floods. A spring brings water to the surface from a subterranean stream, supplying the several hundred residents with water for their irrigated gardens, palms, chickens, and a few goats and sheep.

The people, light-skinned Arabs and dark Sudanese, spoke Arabic, and those who had gone to school spoke some French. They were honest and friendly. We soon learned that we never needed to lock the truck, even when we left valuables such as tools or sunglasses in it.

There were simply no thieves in Béni Abbès, in contrast to what was said about other parts of North Africa.

To the east and rising above Béni Abbès are stunning, golden-red sand dunes that continue without interruption for hundreds of kilometers. To the west, across the wadi, lies the *hammada,* the flat stone desert, stretching far beyond the horizon.

All the houses were built from dried mud bricks, orange-red like the surrounding soil and sand. On a hill above the oasis the striking white-washed buildings of the French military post were visible from far away. A captain, two lieutenants, and a couple of dozen enlisted men acted as the civilian administration, deciding local affairs, sending out rescue parties, and functioning as the highest local authority.

Those in our group who had arrived before Dick and me looked a little pale and weak, victims of more or less severe stomach upsets. The explanation was nearby: swarms of flies congregated in the kitchen,

where they settled on the food. Nothing was done to combat the flies until I sprayed with insecticide and thousands of flies fell surprised to the floor. Evidently these inexperienced flies had not yet acquired insecticide tolerance.

We continued to suffer from occasional bouts of diarrhea as long as we remained in Béni Abbès, although the sanitary situation improved markedly when we moved to our permanent living quarters, a sprawling one-story mud-brick structure assigned to us by the captain. It had been built as a hotel but had never been used and was standing empty. It suited our needs well.

Shortly after we moved in, we were invited to a party by Professor Menchikoff, director of the small research institute in Béni Abbès, who was on a visit from Paris. He had arranged for nomad friends of his to host a feast for all scientists and their families, the French officers, and the local schoolteacher.

The nomads set up their low-slung tents in the open desert within an easy walk of the oasis. We were invited to remove our shoes and sit on rugs that covered the sand. When everybody had arrived, our hosts carried in three large bowls of noodle and mutton soup and large round trays with bread and spoons. A group gathered around each soup bowl, breaking off pieces of bread and eating the delicious soup directly from the common bowl.

Then two whole roasted sheep were carried in, and we tore off large hunks with our bare hands. The next dish was couscous, the most important source of carbohydrates for the Arabs. Prepared from wheat, couscous looks like steamed pearl barley and is traditionally served on large, flat basket trays with a red-pepper sauce that tastes like liquid fire. We again received spoons for eating directly from the common tray.

The meal ended with small glasses of strong, sweet green tea. Our empty glasses were returned to the tray, filled, and redistributed without consideration of who had used the glass before. This procedure was repeated for a mandatory third glass of tea.

After the meal, the nomads offered us camel rides, and I was eager to try. Although I was thoroughly familiar with horses, I had never been close to the huge and ungainly animals that would be the object of our work for the coming year. The camel looked terribly tall; the saddle sat very loosely on top like a small chair. It was much too high to be mounted, but a helpful nomad made a gurgling sound deep in his throat while he jerked at a short rope hanging from the camel's right nostril. The animal swung its head back and forth, moaned and groaned, burped, and bellowed in protest. Suddenly the front legs folded, then the hind legs collapsed like a house of cards, and the camel lay on the ground with all four legs neatly folded under him.

The proper way to mount was to put a foot on the camel's neck before getting into the saddle, but a foot-long vertical stick with side arms like a cross stuck up at the front of the saddle. It was in my way, and I chose a less graceful way to climb into the saddle.

The nomad again jerked at the rope. Evidently getting up was even worse than getting down, for the camel burped and bellowed and belched up a greenish, foul-smelling fluid. At first, nothing happened as I sat balanced on top; then I was thrown forward as the camel stretched its hind legs. Just as I was about to be thrown forward over the camel's neck, the front legs straightened and I sat there, high above the ground.

A nomad held the rope and walked the camel, but I felt unsafe as I swayed from side to side in the flimsy saddle. The cross at the front of the saddle seemed perfect to grasp, but it turned out to be only decorative sticks, so I had nothing but the camel's wool to hold on to. I felt safer when I again had my feet on solid ground.

I looked with admiration at a nomad who jerked at the rope and, when the camel's front legs collapsed, leapt gracefully into the saddle before the camel was completely down. In the same instant the animal was up again and on its way in an easy, undulating gallop. Later, as I became familiar with camels and learned how to handle them, my attitude toward these huge animals improved vastly.

Camels, of course, were why we were in Béni Abbès. We had come to study water balance in these large desert animals, legendary for their ability to go for long periods without water.

In an intensely hot desert, where humans would be close to death after a day without drinking, a camel might last a week or more, depending on factors such as air temperature, how much the camel is working, and what kind of load it carries. We wanted to know what accounted for the animal's extraordinary ability to survive. We had seen how kangaroo rats adapted to life in the Arizona deserts; by what mechanisms did camels, roughly ten thousand times the size of kangaroo rats, adapt to the Sahara? Whereas kangaroo rats escape the heat of the desert day by staying in their burrows, camels are exposed to the full heat of the desert sun.

One challenge was to determine whether camels rely on evaporation to keep the body temperature from rising unduly, similar to the way humans sweat to prevent overheating. Common sense told us that camels must evaporate water when it gets truly hot; otherwise they would die from overheating. But would they sweat like humans or pant like dogs to keep cool in the heat?

When the air temperature is higher than the body temperature, as it often is during the summer months in North Africa, camels need to drink—but how much? Have they somehow evolved mechanisms to reduce the amount of water needed to keep cool? In hot deserts, such as Death Valley or the Sahara, a human will be close to death after one day without drinking. Desert travelers have reported that camels can do much better, but how do they do it? Might camels tolerate losing more water per unit volume than other mammals? Or might they store water in their bodies, drawing on those reserves when faced with dehydration? But if so, where? Anecdotal accounts maintained that water is stored in the camel's hump, which actually is filled with fat as an energy reserve. Other reports, dating back two millennia to the writings of the Roman naturalist Pliny, claimed that camels store water in their stomachs. We doubted this, but we needed to rule it out with certainty and also to find out whether water could be stored elsewhere in the camel's body.

We also needed to know how camels regulate their body temperature on a scorching hot day. Humans have a narrow range of body temperature, around 37°C (98°F). In a hot desert we maintain this temperature by sweating profusely. Might camels use less water if their body temperature were permitted to rise, rather than remaining constant? And if so, how high a body temperature could they tolerate?

Finally, we wanted to know if camels could tolerate a greater loss of body water than other mammals and thus be more tolerant of dehydration.

Before we could begin our study, we needed a place to set up our laboratory equipment and somewhere to keep camels. The captain at the French military post assigned us a large abandoned building a few hundred meters above the village on the hammada. The barren yard, surrounded by a tall mud-brick wall, was ideally suited for our studies. The building even had an archway about four meters high—high enough to attach the scale for weighing camels and to suspend them clear off the ground, which required cooperation from the animals. Eventually, by proceeding slowly and gently, I succeeded in weighing all the camels we studied.

Finding suitable camels for sale proved surprisingly difficult; animals the natives would sell were either too old, too expensive, or too sick. Fortunately, the captain had assigned us an assistant, a young man named Mohammed ben Fredj, who became our indispensable helper and friend. He spoke excellent French, an essential asset because I knew only a few phrases in Arabic. Mohammed had fine features and dark skin like his mother, who was of Sudanese origin; his father was a nomad of the Chamba tribe. Professor Menchikoff told me Mohammed was twenty-six or twenty-seven years old, but he looked more like twenty. He was totally reliable and equally at ease with Europeans and local people.

I had hoped to buy only female camels. Some male camels were available, but males are less pleasant than females. Work camels are castrated, but even these are more difficult than females. In addition to their ease of handling, females are easy to catheterize, which is necessary for studies of kidney function. Male camels have an S-shaped urethra, making catheterization hopeless. The nomads, however, prefer to keep their females for breeding, which made our search for suitable animals difficult.

I liked a sixteen-year-old female that Mohammed had located. She was quiet and friendly, but the owner asked an unreasonable 40,000 francs (about $120), and the camel was too old anyway. Mohammed said that considering her age about 25,000 francs would have been normal, and at that price I might have bought her in spite of her age.

We finally located a beautiful, slender four-year-old female that had not yet reached full adult size. A camel is mature at five, is in its prime at seven or eight, and may live to age fifteen or twenty. She had been treated well, as is usual among the nomads, and was in excellent shape. The owner asked 28,000 francs for the female and would not settle for less. Mohammed and I dickered unsuccessfully with him for a couple of hours, but when I gave in, he suddenly raised the price. I merely refused to pay more and handed him the agreed amount.

This animal rapidly became tame, and I often gave her dates, which she very much liked. As the year went by, I often enjoyed spending time with her and became more attached to her than to any of our other camels.

One morning Mohammed told me about an adult female that the owner was asking 34,000 francs for, so I went to look at her. She lay quietly while we examined her; she had no obvious signs of disease and no fever to indicate trypanosomiasis, a common blood parasite. She was lean but in good shape, with a well-developed and firm hump, a sign of a good nutritional state. The owner said she was eight or nine years old. We sat for hours talking about the price, which gradually came down to 30,000, which was cheap for a fully grown, healthy female. I asked if she was pregnant, for that would make our work difficult. The owner said that he didn't know, but that if she was, she would be in the second month, too early for us to diagnose. Pregnant females are valuable, and we concluded that the owner was willing to sell because he knew her to be sterile.

This camel had never carried a halter and might be difficult to manage, so I offered the owner 1,000 francs to stay for another day to help get her used to a halter and rope. No, he said, he could not possibly stay; he had to leave immediately. I paid the 30,000 and the seller left.

The next day, when she refused dates, I began to suspect that something was wrong. Why had the price been so low and the owner so anxious to leave? Even if he needed money, he would not have sold a pregnant female because by next winter he would have two camels. She could be sterile. Or the poor appetite might indicate illness, but there was no obvious sign of disease. After a few days Mohammed asked another nomad to look at the camel. She was pregnant, the nomad reported, and very old. Her teeth were so badly worn that she could not eat dates even if she wanted to. The dates used for camel feed are different from the soft and sweet dates we eat—hard, fibrous, and very bitter. In our excitement at finding a healthy female, we had forgotten to look in her mouth! At her advanced age pregnancy would probably kill both mother and calf, the nomad said, so I was the owner of an old, pregnant camel that was of no use to us. Sadly, we had to butcher the animal and sell the meat to recover some of our loss.

Béni Abbès turned out to be an ideal location for our work. The climate was right, and the camels we bought were acclimatized to desert conditions. Considering the circumstances, we had excellent working conditions. Our building was an empty shell that lacked heat, water, and electricity, but it met our needs, and the surroundings were gloriously beautiful.

To begin our studies, we had to weigh each camel to establish a baseline from which we could calculate water loss. However, you can't just lead a camel to the hoist and expect cooperation. A camel that isn't used to being handled is frightened of anything new, and approaching it with a huge piece of canvas terrifies it. Imagine how a horse that has never been handled would react if suspended in a canvas sling with the legs dangling free. However, camels are basically quiet and friendly, and with patience and an occasional date, I succeeded in leading the animals to the hoist for weighing. While Mohammed distracted the animal with a few dates, I could secure the canvas sling around it and slowly hoist it up and read its weight on the balance.

The weight loss from day to day would give the total water loss, except for the weight of the small amount of feed they ate, some dates and dry hay, which we carefully monitored. The dates were hard and tough, and the hay was merely tufts of dry grass, hand picked among the sand dunes and bought for a few francs.

Much of our own food came from Colomb Béchar, 250 kilometers and at least five hours' drive north on the miserable roads. During the winter months we could arrange with a store to send us food packages on the weekly bus, but when the weather turned hot, most food items could no longer be sent by bus, and we were limited to what we could get in Béni Abbès.

The "bus" was a huge truck that ran south through the Sahara for more than 2,000 kilometers, from Colomb Béchar to Gao, east of Timbuktu. Béni Abbès was a short side trip on this long run. Several passengers usually squeezed into the cab with the driver, while others sat with their bundles on top of the load of boxes, packages, and spare parts for broken-down vehicles. Sometimes the truck broke down and was delayed for days, and occasionally a flood in the wadi stopped all traffic.

On our first shopping trip in our own truck we started out in predawn darkness at four-thirty and arrived at Colomb Béchar at ten. Our first stop was the outdoor market, where we bought some beautiful tomatoes that were simply handed to us. Nothing was put in bags, and with our hands full of ripe tomatoes, we learned that a basket was as essential as money for shopping. We bought a large basket woven from palm leaves, picked up our tomatoes, and continued shopping, dumping everything we bought on top of what was already in the basket. Later, when we removed our purchases, we learned that one should buy ripe tomatoes after the cabbage.

At noon, long before we had accomplished what we had planned, all stores closed for a three-hour siesta. We sought refuge and lunch in the local hotel. When the shops reopened, we found an excellent food store that carried items we had never expected to find—Danish butter and dry milk powder, French paté, and good cheese. We accomplished a great deal between three and six, when it turned dark and we started home, arriving at eleven.

Our usual breakfast was bread, butter, and marmalade, with coffee or cocoa made with dry milk. Lunch was fresh bread, butter, sometimes eggs, canned liver paté or sardines, and cheese; sometimes our cook, Lagrib, prepared a warm dish. If a goat had been butchered, dinner would consist of meat, which Lagrib served with potatoes or rice, perhaps a vegetable, sometimes couscous. We had camel meat only a few times because camels were rarely butchered. During the first few weeks we often had oranges for dessert, but by December oranges and most vegetables had disappeared, as had eggs; the hens laid none during the winter months. However, we always had fresh bread, baked locally.

In the laboratory we worked a rigorous daily schedule. Stig, Dick, Bodil, and I agreed on our research protocol and divided the labor among us. I weighed the camels and took a major role in deciding how much water and food to give them and when. The others did more of the "bench work," or analytical work, each taking a particular method to perform regularly, except for a complex method for determining blood volumes, which I knew better.

Dick or I measured the camels' rectal temperature, and when we needed twenty-four-hour measurements, I stayed with the animals all night. At times I worked at a small table in our living quarters, keeping accounts for our shared living expenses and corresponding with Duke about timely payments from our grant funds. Bodil seemed to resent the time I needed to keep everything running smoothly, but as a professional team the four of us worked well together despite occasional marital strife.

As our work progressed, our results were increasingly exciting. As we suspected, our measurements showed that on a hot day, a camel that needs to conserve water permits its body temperature to increase by several degrees. Indeed, a camel's body temperature may drift up to 41°C (106°F), thus saving water that would otherwise be secreted as sweat to cool the body. It was obvious that a temperature of 41°C in a perfectly healthy animal in no way indicated "fever" as customarily defined.

The increase in body temperature during the day means that heat is stored in the camel's body, and during the cool night, when the air temperature drops, the camel can cool off without using water. In addition, if body temperature drops below normal at night, the animal can tolerate more heat exposure during the next day. We found that if the camel's morning temperature had fallen as low as 34°C (93°F), it would take longer for its body temperature to reach 41°C, the limit that a camel can tolerate. To prevent an even higher temperature, the camel must begin to sweat.

We had seen several reports that camels do not sweat and lack sweat glands, but my suspicions that camels do indeed sweat were subsequently confirmed when I shaved a small area of skin and on a hot summer day saw tiny droplets of sweat. Later, microscopic examination of skin samples revealed numerous sweat glands. However, our camels never sweated profusely, and in the dry desert air the moisture evaporated before the fur became visibly damp.

An increase in body temperature has one advantage that may not be immediately obvious. Say that on a very hot day the air temperature reaches 45°C (113°F). It takes more water to keep a body eight degrees below the air temperature, as humans do, than four degrees cooler, as camels do. Air that is warmer than the body tends to heat an animal because heat always flows from a higher to a lower temperature, and in the desert the situation is exacerbated by intense radiation from the sun. It turns out that camels have yet another line of defense: their fur is a good insulator and blocks the penetration of heat, which in turn means that less water is needed for cooling.

An obvious way to test the role of the fur in heat regulation is to shear a camel and see if its use of water increases. Not surprisingly, when a camel is shorn in summer, its use of water increases by about 50 percent. It may seem counterintuitive that insulation is an advantage in a hot climate, but measurements on humans show that we sweat less on a hot desert day if covered with loose clothing. The fur of the camel similarly reduces the heat load and the use of water.

The camel has one further advantage: it can tolerate a greater degree of dehydration than other mammals. Humans or dogs are close to the lethal limit if they lose 10–12 percent of their body water. A camel tolerates losing twice as much, perhaps 20–25 percent. We do not know the exact limit, for no sane person would make the determination by depriving a camel of water until it dies from thirst. However, we can assume that in an intensely hot desert, where a human would be close to death after a day without drinking, a camel might last for perhaps a week or more.

When a camel is severely water depleted, it drinks enormous amounts to restore water balance. Some of our camels drank over 30 percent of their body weight within minutes. One young female weighed 201.5 kilograms (444 pounds) when she started drinking; 10 minutes later she had drunk 66.5 liters (more than 17 gallons) and had increased her weight by 33.1 percent. If other animals were to drink in a short time proportionally as much as a camel does, they would die from water intoxication.

Travelers who see a thirsty camel drink may conclude that the animal is filling up with water, stockpiling it for future use. Yet our evidence provided no support for this old belief. When deprived of water for some time, our camels drank enough to regain water balance, but not in excess of what was needed, and I am inclined not to believe in water storage. Still, the issue isn't fully resolved and deserves further study.

The problems we studied changed with the seasons. We had to wait for summer to obtain information about the camels' reactions to extreme heat. In winter there was a great deal to learn about other questions, such as their need for water when there is no heat stress; we also had to establish baselines for their normal physiology. In the meantime, the routine of our daily life in the exotic surroundings continued, with occasional surprises.

In mid-January I saw a few locusts settling on the ground; then more came flying and alit here and there. Suddenly a cloud of locusts came, all from the same direction, rattling like dry leaves as they came down everywhere, chewing up every leaf they landed on. Yet this swarm was nothing in comparison with a true locust plague, when a solid rain of locusts continues for days, devouring everything green and leaving nothing but stripped trees and denuded earth.

The next day I had tea at Mohammed's house, and we discussed the locusts. "People eat them," he told me. "They collect as many as they can, and what they don't eat immediately they save for later." When I asked how people prepared them, Mohammed explained: "First they are boiled; then people break off the wings, the legs, and the antennae. You can eat the boiled locusts immediately, but they taste better if they are fried in oil."

When I said I wanted to taste them, Mohammed promised to bring us some that evening. I told Lagrib merely to put them on the lunch table with the other food the next day without commenting. Nobody found it unusual that we tried new kinds of food, but the locusts weren't popular. It was like eating tasteless shrimp without peeling them. Chewing thoroughly didn't help because the pieces of hard cuticle were difficult to swallow. The experience was interesting rather than enjoyable, and we never tried locusts again.

One day in the spring we had *dob*, the Arabic word for a lizard whose scientific name is *Uromastix*, literally "fat tail." Aptly named, these plump, grotesque animals have broad, scaly tails. The only reptiles I had eaten before were turtles and rattlesnakes, but because Mohammed said the Arabs like dob, I wanted to try them too. I obtained four large specimens and asked Lagrib to cook them. He clearly didn't like the idea. "We could start with one," he suggested. He was not used to Europeans eating native food, and he feared we wouldn't like what he served; then he would be blamed for the failure, and thus disgraced. However, with no other instructions for dinner and no other food to cook, he reluctantly prepared the dob. Everybody liked the result. The texture was similar to that of chicken, the taste was excellent, and the liver reminded me of chicken liver. In fact, dob was far better than the tough hens in Béni Abbès, which had a strong, strange flavor.

Toward summer our food became less varied. There was no more meat, no butter, no cheese, and very little else. The selection of vegetables was limited; little was grown in Béni Abbès. We never again had oranges, but we could always get dates. Somehow, dates are less tasty when they make up a major part of the diet. Bent, who never liked sweet things, soon tired of them, as did most of us. I ate dates longer than the others, but when it turned really hot I couldn't stand them.

Béni Abbès had excellent water. It was kept in *goulas,* large clay jugs in which evaporation from the porous surface kept the water cool. When the weather turned hot, Aïcha, our maid, put bottles of water in our bedrooms, but it was always lukewarm and as refreshing as drinking from a hot-water faucet.

As we adjusted to local customs, our view of the need for modern hygiene changed. During a visit to Mohammed's house he offered us dates and sour goat's milk. We knew the milk was kept in a *gherba,* a bag of goatskin covered with black fur, yet without a thought we drank it from the same bowl as Mohammed.

Because gherbas are light and don't break, they are always used to carry water for desert travel, as we learned on a visit to the nomads. In early summer, when we knew them well, the nomads invited us to visit them in the desert. They had moved their camp from the stone desert, the hammada, to the sand desert, the *erg.* They came with camels to fetch us for a five-hour journey in an infinity of burning sand with the fiery sun overhead. The endless sand dunes were awesome, and we were only at the edge of the Grand Erg Occidental. After a couple of hours in the heat, alternating between walking and riding a camel, I found the cool water in the gherba the most refreshing drink I had ever tasted. Every drop was precious, and the tarry taste and the floating black particles in the water were of no concern. I developed a positive fondness for the peculiar black furred bag that dangled against the flank of the camel.

The children, who had their own routine, accepted life in the desert as the most natural thing in the world. Because a year without school would otherwise have put them behind on their return to Durham, Bodil had arranged a correspondence course for them, and before we left Durham we received a year's supply of teaching materials. Helga taught Astrid and Bent, and Doris Houpt took care of Mimi's schooling. Bent and Astrid also attended the French school with the local children and soon spoke fluent French. Bent found a good friend in Lagrib's son, who was about the same age. They spent a great deal of time together, doing whatever nine-year-old boys do in the desert to entertain themselves and mystify their parents.

In winter the sun set shortly after five, and night descended rapidly. With no electricity, if we worked as late as six we needed a flashlight to negotiate the steep, stony path home. I also provided everyone with a flashlight to move about indoors.

We needed light in the kitchen and for reading in the evening, but candles were expensive and gave inadequate light. I bought a kerosene lamp, which was adequate for the dinner table and to read and work by. Lagrib needed better light for cooking and for dishwashing, so I bought a carbide lantern for the kitchen. It gave a bright, radiant light, and dishwashing improved visibly; now only minor amounts of food stuck to the plates. Carbide lamps are tricky, and I enjoyed the challenge of keeping ours working. I emptied and cleaned it every day, recalling the idiosyncrasies of the carbide lamps I had used years earlier as a miner in Norway.

I often woke before five and heard the cocks crow, but in Béni Abbès the cocks crowed all night and were useful for nothing but dinner. I found it boring to lie awake in total darkness, unable to read, so I usually got up and spent an hour or two typing letters in the light of a small oil lamp I had bought in Casablanca. For a wick I used a piece of cloth Mimi gave me, and when I filled the lamp with olive oil, it burned with a small, odorless flame. The Arab who sold it to me had said that the lamp was old and came from the mountains. It was made of a soft, light gray soapstone that could be worked with simple tools. The shape was classical, with the neck to one side and a wick at the tip. The stone around the wick was pitted from years of use, and I wondered how old the lamp really was. As soon as oil touched the lamp, the stone turned nearly black, which convinced me the lamp hadn't been lit for hundreds or even thousands of years. It gave me a sense of satisfaction that I had filled it with olive oil, much as its previous owner had done long, long ago. On bitterly cold winter mornings, when the temperature hovered close to freezing, I got thoroughly chilled when I sat typing in the dark house. The oil lamp, sitting on the table next to my typewriter, put out a meager amount of light, but it lent a little warmth to my icy right hand.

One of my most memorable encounters with nighttime temperatures took place toward the end of January, when we drove to the nomad camp to buy a sheep. I timed our visit so we would arrive in the afternoon and avoid the big meal that would undoubtedly be offered us by explaining that we had to return home before dark. As expected, the nomads were happy to see us and immediately invited us to stay overnight. I told them we had work waiting at home, animals to be weighed and fed, and, most important, we were expected back.

The chief, the *khaïd,* did not accept our refusal. He reminded me not only that I had excused myself the two last times I had visited him, but also that I had promised to return to be their overnight guest. Now he was obliged to conclude that I did not care to be his guest.

How could I politely explain our predicament? If we failed to return, there would be an alarm and then a search. However, nobody knew exactly where we were, for the nomads had recently moved their tents, and the only man who knew their new location had come with us to show us the way. We always carried food and water in the truck, but there would nevertheless be concern, and the captain would be asked for help.

The Arabs looked grim when I refused their hospitality, and after some embarrassing minutes I opted to become hostage to diplomacy: I would remain while the others returned home, to come back for me in the morning. However, before they were allowed to leave, they had to drink the obligatory three glasses of sweet green tea, then eat roasted cubes of liver wrapped in pieces of omentum. It was then I realized that, as soon as we had arrived, a sheep had been butchered and a feast planned.

After Mohammed and the others left, communication with my hosts was reduced to a minimum. The nomads knew only single French words, such as *bonjour* and *ça va,* and because there are limits to how many times one can repeat greetings, I was restricted to sitting and eating while the nomads carried on a lively conversation among themselves. When the couscous and roasted sheep were carried in, each person got a spoon and made a depression in the couscous for the sauce, thick with *fil-fil,* the powdered red pepper that burns like fire. The khaïd was especially polite to me, his honored guest, and over my vigorous protests ladled generous amounts of fil-fil onto my couscous. After the lengthy meal we had the obligatory three glasses of super-sweet tea, nauseating after the excess of meat and couscous I had eaten, but unthinkable to decline. The nomads sat for hours and talked over the tea, while I got colder and colder in my thin trousers, short-sleeved shirt, and bare feet in sandals.

The guests gradually vanished. My hosts gave me a thin blanket and a place to sleep between two of eight or ten nomads. The tent was open along one side, but luckily there was no wind. I slept very little as the cold slowly penetrated my body, and I lay there wondering if I would ever be warm again.

When morning finally came, I got up, stiff and nearly unable to move. The ground and the tent were covered with frost. The nomads made a small fire, barely big enough to heat water for the obligatory three glasses of tea, for fuel is at a premium in the desert and is never wasted. By midmorning the truck came to pick me up, and only then was I allowed to buy the sheep we had come for the day before. Later I learned that it had snowed in Colomb Béchar and the low temperature was a record.

Winter was the time for rain. In a normal winter there might be a single shower or no rain at all, but while we were in Béni Abbès it rained several times, beginning in late November.

In December we had a great thunderstorm with a brief downpour. The road to Colomb Béchar was too soft for traffic and was closed. Rain was always welcome, but the houses were not built to withstand it. The flat roofs were supported by horizontal palm trunks, overlaid with stems from palm leaves, then layers of palm leaves with the fronds, and finally a heavy layer of mud.

Most of the year this construction was fine, as the palm leaves kept the dry mud from falling through. But when the first major thunderstorm came, our delight lasted only a few minutes, for muddy water started coming through the ceiling and formed large puddles on the floor. After the rain, wide orange-brown stripes formed interesting patterns on the whitewashed walls. With the next downpour, torrents came through the roof. Every small crack was rapidly enlarged by the streams of water, and the tin cans and pots we put in the worst places rapidly filled. We ran around and emptied them into larger pails, then emptied these outside. Lagrib went up on the roof and tried to seal the worst leaks with mud—a hopeless job while the rain poured down and water flowed everywhere.

The gray, chilly weather continued, reminding me of cold and rainy autumn days in Norway. Luckily, the showers were brief, and as long as there were no showers at night, things weren't too bad. It rained at night only once, and not very heavily, so we could go back to sleep after moving our beds out of the way of the dripping water.

The rains that fell in Béni Abbès were never sufficient to flood the wadi. We were in a true desert, and all moisture was soaked up by the thirsty ground. However, a few times a telegram to the captain warned him that heavy rains in the Atlas Mountains had caused floods that would reach us in a few days. Then masses of water at the speed of a fast walk came foaming in a wave down the dry riverbed, spreading out and continuing to rise for hours, forming a raging stream. The floods lasted for days while the water slowly receded. Then we received no mail, no food packages, nothing, until the bus could again cross the river.

I WAS INTERESTED in the several kinds of animals that lived in the desert and often thought about possible later studies of their water metabolism. The French zoologist Francis Petter, who was visiting Béni Abbès that spring, told me about fennecs, diminutive desert foxes that grow to about half the size of a cat. They live among the sand dunes and have remarkably large ears that stand straight up. Fennecs are said never to drink water. They eat small animals, rodents, lizards, and all sorts of insects, as well as some plant material and berries. In April some people in Béni Abbès dug out fennec pups and took them home to fatten them in their households to eat or sell.

Petter and his wife, a geologist, told me that perhaps one out of ten fennec pups grows up to be friendly and easy to handle, while the rest will be ill-tempered and difficult. As small pups, however, 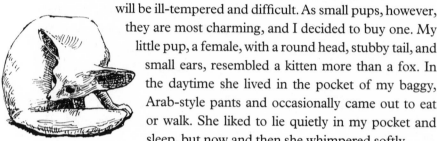 they are most charming, and I decided to buy one. My little pup, a female, with a round head, stubby tail, and small ears, resembled a kitten more than a fox. In the daytime she lived in the pocket of my baggy, Arab-style pants and occasionally came out to eat or walk. She liked to lie quietly in my pocket and sleep, but now and then she whimpered softly.

At first the fennec was not used to my smell, and if I held her in my cupped hands and breathed gently into her face, she bit my hand and yelped, sounding precisely like the distant bark of an adult fox. Once she knew me, however, she snuggled farther down when I breathed softly over her.

At night I kept the fennec in the bedroom, where she slept in a basket with my pocket handkerchief to make her feel at home. If she woke and whimpered, I got condensed milk for her, just as if I had a baby in the house again. When she cried, I took her into bed with me, where she slept quietly in the corner of my arm.

I wasn't too happy the day she had diarrhea in my pocket and her liquid feces ran down my leg, but I changed my clothes and allowed her back. When she grew bigger I gave her a piece of liver I had removed from a dead mouse, putting it in her mouth. At first she chewed automatically; then she suddenly metamorphosed into a fierce little carnivore. She growled and snarled and tore at the carcass, arched her back, and bristled if I came near. When I gave her a second piece of liver, she was so eager that she nicked my finger with her small, sharp teeth. One taste had transformed her into a beast of prey.

Eventually she moved from my pocket and slept quietly on my lap, but as she grew she became more active. On one occasion I was lying on my bed reading and eating a piece of chocolate with the fennec dashing about. She crawled over me, tried to burrow under me, and came eagerly up to my face as if to creep into my mouth; she sniffed and bit me gently on the neck and fingers. Thinking that the chocolate interested her, I gave her a tiny piece. She chewed it eagerly but did not growl or act like a carnivore, only showing a great enthusiasm for chocolate. After giving her another piece, I stopped, fearing the return of her diarrhea.

Petter later brought us a second fennec, a male, and we later took both of them back to Durham. After the male's arrival the female grew less tame with me; she no longer snuggled up to me as she had when living in my pocket in Béni Abbès. The fennecs lived on an enclosed porch and ate table scraps; they were especially fond of small pieces of peaches and strawberries from a local garden. We bought nice-looking strawberries in a store, but the fennecs refused to touch them. I assumed that the commercial strawberries had been chemically treated to reduce spoilage, with results totally objectionable to the fennecs though not noticeable to us.

In Béni Abbès, at Petter's urging, we kept a hedgehog as a form of pest control. During the day it slept, and at night it pattered softly around in our rooms, eating insects and scorpions. Every evening we put the hedgehog in Bent and Mimi's room, where it rummaged around for a while before squeezing through a big crack under the door. It then ran back and forth at the two steps in the hall, not daring to jump. At last we would hear it go down with a light thump, then another, and the animal would come to our room. Then we heard the soft patter of its feet and the gnashing of teeth when it found something edible. It went around like a tiny vacuum cleaner and absorbed whatever edible tidbit it came across.

Petter gave us a second hedgehog, which attached itself to the Jarnums. This hedgehog loved to climb, and each night it ascended into Mrs. Jarnum's bed, where it seemed thoroughly satisfied to sleep. Mrs. Jarnum was not entirely happy with this arrangement and soon learned to be careful before turning in her bed.

Eventually we got a third hedgehog, which became Astrid's. Astrid once forgot to close the door to her room, and her hedgehog disappeared. The next day a man sold me a suspiciously tame hedgehog that was exactly the same color as Astrid's. When I took it home, it immediately crept into the special place that Astrid's old hedgehog had created in her wardrobe. Although I was certain I had bought back our own animal, the next morning Astrid said that her old hedgehog had had different nocturnal habits. Petter agreed with Astrid, saying the animal he had given us had a small spot where the tips of the spines had been crushed, and the new hedgehog didn't have this spot. In any case, Astrid liked the new one and was happy to share her room with it. More than a week later, Lagrib found the old hedgehog in a small room on the other side of our courtyard. Astrid had been right all along.

In the laboratory we had six jerboas, but we didn't divert time from the camel work to study their water metabolism. Although they belong to a different family of rodents, jerboas are amazingly similar to kangaroo rats: they have long hind legs for hopping, short front legs for digging, and a long tail for balance and steering; they are nocturnal, live in burrows, eat seeds and other dry plant material, and never drink.

We also had gerbils, which, like jerboas, got along well on dry seeds and no drinking water. The gerbils ate peanuts, which neither kangaroo rats nor jerboas could survive on because the high protein content produced a need for extra water. The gerbils could also tolerate eating *Salicornia*, a plant with a salt content about twice as high as that of sea water. Gerbils are definitely the champions of desert survival.

W$_{\text{ITH THE EXCEPTION}}$ of occasional visits from travelers, our social life was relatively restricted. A few times Mohammed invited Bodil and me to his house. He always served us the traditional three glasses of green tea, sweetened with immense amounts of sugar. His house was on the steep, rocky slope toward the hammada, above most of the other buildings. To reach it we walked up a narrow passageway between other houses and climbed steep steps to enter a small, square yard surrounded by tall walls.

Mohammed's room opened onto the yard. Inside he had many rugs on the floor and a small radio, but no other furniture. The central steps in the yard led to more rooms on other levels, and a stairway went to the roof above Mohammed's room. Behind, a steep cliff continued up to the hammada, leaving room only for a tiny garden with a few palms and a couple of fruit trees.

It was obvious that Mohammed was well-to-do. To the right was the entrance to a cave, and Mohammed explained that he slept there when it was too hot, especially during Ramadan, the ninth month of the Islamic year. This occasion is marked by strict fasting from sunrise to sunset. Neither food nor drink is allowed until it is so dark that a white woolen thread can no longer be distinguished from a black thread.

By our calendar the time for Ramadan changes from year to year, and that year it came in June, when the temperature might soar to 45°C (113°F). It was very hot, and drinking nothing from five in the morning until after sundown was rough, so I suggested that Mohammed leave work at eleven and return at four. In this way he could sleep in the cool cave during the worst heat.

On July 2 Ramadan ended with a big celebration near the piscine. A man read from the Koran; then all the men faced Mecca and prayed in unison, first standing up, then kneeling with foreheads against the ground, then standing again. With fasting over, Mohammed resumed a full workday, and Lagrib recovered his pleasant disposition.

One day Mohammed respectfully asked permission from the khaïd of Béni Abbès to take us through the *ksah*, the original Arab housing area. "If one doesn't ask permission, somebody might be offended," Mohammed explained, "but if one has asked, there will be no criticism."

In the ksah the houses are built together to form a huge continuous structure with the passages between them roofed over. With Mohammed leading, we walked in total darkness, feeling our way along the walls, often not knowing whether we were inside or outside a house. Without Mohammed we would have been hopelessly lost in the dark maze.

In a passage near the mosque, a blind man measured the water that flowed in a narrow channel from the piscine to irrigate the gardens. He sat near a large container of water into which he placed a copper bowl with a small hole in the bottom. The water streamed into the bowl, which gradually filled and sank to the bottom with a light clank. When the man heard this, he lifted the bowl and started over again, making a knot on a thin strip of a palm leaf. After a predetermined number of knots, another man went to redirect the water to the next garden.

While we were with the blind man, an older Arab invited us to tea, and we gladly accepted. The tea was excellent, less sweet than usual, and flavored with fresh mint. The man was the khaïd's cousin, Mohammed explained, and his house was built as a continuation of the mosque. The entry hall of his house had a hole in the wall where we could look down into the mosque.

After tea Mohammed said that we could see the mosque, although earlier he had told us that we could not even approach the door. The khaïd's cousin evidently had authority to permit infidels to see the holy place. The mosque was a simple and beautiful structure with clay brick columns and arches, plain whitewashed walls, and a bare earth floor that in winter would be covered by rugs. There was no need for seats because prayers are said either standing or kneeling.

Two burial coffins hung on the wall beside the door. Each was a stretcher, coarsely made from unpainted wood, used only to carry the dead to the grave and then returned to the mosque. One coffin had a cover and was for women; the open one was for men. Even in death a woman must not be seen by a stranger.

This was not the first time I reflected on the custom of wearing a veil. Among the Tuaregs, a tribe in the Hoggar Mountains of the Central Sahara, men wear veils and women do not. Are these men as embarrassed about showing their faces as the women were in Béni Abbès?

As a child I had learned in school about Kemal Atatürk, the first president of modern Turkey, and the freedom Turkish women felt when allowed to remove their veils. But in Béni Abbès I understood that a woman who has always worn a veil feels most uncomfortable if required to reveal a part of her body that only her husband is supposed to see.

As a student I had read a memoir by a noted Danish archaeologist, Valdemar Poulsen, who attended a diplomatic reception in Ankara in the early 1920s, shortly after Kemal Atatürk's vigorous modernization program began. Professor Poulsen noted that the women, who appeared in public without a veil for the first time in their lives, looked as embarrassed as any Western woman would had she been required to attend a diplomatic reception with her breasts bared.

On one cold spring day the wind increased steadily, rather than dying down toward evening as usual. Violent gusts whipped sand into my face and eyes, and I could hardly see where I stepped. The children covered their heads with gauze and ran in and out of the house, thinking it great fun.

I had read that the light during a sandstorm is yellow, but I found that it was a peculiar, unreal gray, brighter than dusk but dimmer than normal daylight. Occasionally I saw a glimpse of the sun as a pale disk through the clouds of sand, like a weak white moon.

Sand came in around the windows and through keyholes, penetrating every little crack in the house and settling everywhere. Sand penetrated our clothes, and our cameras were ruined unless kept in plastic bags. Despite my efforts, only one of my cameras survived.

The sandstorms made me recall an English soldier from the Foreign Legion whom Dick and I had met in Saïda on our way to Béni Abbès. We had stopped for repairs and were eating dinner at the hotel when two legionnaires, a young German and an Englishman, stopped to talk with us. The German soon left, but the Englishman said that he would enjoy talking with us further because there were so few English-speaking people in the Legion. I bought him some wine, and he started talking about himself. Before joining the Legion he had worked for seven years as a chauffeur on several expeditions with the English scientist and explorer Ralph Bagnold, author of the well-known book *The Physics of Blown Sand and Desert Dunes*. The legionnaire initially claimed that he was in love with deserts and had enrolled in the Legion merely to see more of the desert, but after more wine he revealed that he had enrolled in Marseilles, where he was stranded without a penny in his pocket. Bagnold, he told us, could not enjoy a meal unless sand crunched between his teeth. When he returned to London, his club kept on his table a salt shaker with sand to sprinkle on his food to remind him of the desert.

I wasn't that enthusiastic about sand, which always stuck to the dates and was a constant ingredient in the meals that Lagrib cooked. I could tolerate sand in my food if I had to, but I had no desire to maintain the tradition once I left Béni Abbès.

In May the temperature climbed above 38°C (100°F) every day. Swimming in the piscine, which was reserved for Europeans, gave us a welcome relief from the incessant heat. At night the house stayed warm, and we slept on the flat roof, where it was pleasantly cool. The heat radiated to the dark night sky, which acts as a giant hemispherical cooling surface, the low air humidity facilitating radiation loss. Together these factors made the night sky an effective heat sink.

By late May the heat became so oppressive that we decided to send the children to Denmark with Helga. One evening shortly before their departure I was lying on the roof, talking with Astrid and watching the first stars come out. She surprised me by saying that she didn't see the stars as points of light, as I saw them, but as star shapes like the drawings in books. Until then I had no idea she was nearsighted. I wonder how many thirteen-year-olds have had their myopia diagnosed while lying on a roof in the Sahara Desert.

In June the nights were still tolerable, but afternoon temperatures soared to above 42°C (107°F). Indoors the heat was not yet unbearable, but all objects were hot to the touch. Our food, papers, and books were warm. If I sat down on a metal chair, it burned my skin. The only cool thing was the water in the goulas.

Fortunately, the rooftop was still comfortable at night, as long as we were lying down and could feel the coolness as our body heat radiated out to the night sky. However, if we sat up, the hot air swept around us like a warm blanket, and the pleasant feeling of coolness disappeared because less of the body surface could radiate to the sky. If a desert wind came up, it felt like a blast of hot air from an oven.

We went to sleep at dark, at about eight, and after sleeping for a few hours on a still night, we would get so cold that we had to cover ourselves with a sheet. Then we slept peacefully under the stars until the first obnoxious flies arrived at four-thirty. They settled on our faces, preferably on the lips or in the nose, where they found a trace of moisture. Pulling a sheet over our heads offered no protection; they always found their way in. The flies' arrival marked the end of sleep, and trying to sleep inside was useless; just lying still made the sweat run.

Toward the end of June, when the daily temperature reached 43°C (110°F), the heat became truly oppressive. The heat wore us down, and we were looking forward to returning to a climate where the wind blows cool. I was happy our stay was coming to an end. It had been a wonderfully rewarding and memorable year for all of us, highlighted by the success of our research. However, I had developed a distressing feeling of being an intruder in a society where we did not belong. We had experienced a most gracious hospitality, but in the self-sufficient life of the oasis, we were outsiders.

By now our regular diet consisted of dry bread and little else. Even freshly baked bread dried out in a couple of hours. We no longer got food packages from Colomb Béchar. Lagrib could again buy eggs, but it was so hot we had to eat them immediately; if we kept them for a day, a chick promptly started developing. Lagrib had an ingenious approach to the problem of getting fresh eggs: if a woman had more than two eggs for sale, he didn't buy any, but if she had only one or two, he felt reasonably certain that they had been laid the same day. Usually he was right. Meat was even more of a problem. We had no refrigerator, and the temperature was perfect for bacterial growth. Once we tried to keep fresh meat overnight; by morning it had turned green and slimy and foul-smelling.

Bodil and the Houpts left in late June, and I stayed behind with Stig to finish up our last experiments and pack our equipment. Temperatures were now hovering around 45°C, and the nights were intolerable unless the air was perfectly still.

On the Fourth of July we had a sandstorm, with winds blowing continuously into the night. I took a bucket of water with me to the roof, moistened myself all over, and went to sleep. Later in the night, half asleep, I wetted myself down again, then pulled a wet sheet over myself. It dried so fast, however, that I constantly had to repeat the procedure. With one hand in the water bucket I slept fitfully until the flies arrived at the first sign of daybreak.

Stig and I spent the last few days packing and getting ready for departure. By now even the piscine provided only momentary relief: as soon as we were out of the water, everything was burning hot. Moreover, we still had work to do and hated the couple of minutes' walk in the scorching sun back to the overheated laboratory.

Stig and I planned to drive to Algiers, ship our crates and trunks to the United States, and continue north through Europe to join our families in Denmark. But we faced a hazardous journey across the Atlas Mountains on a road that Dick Houpt and I had found nearly impassable on our way south. How would we make it with a truck that was having frequent breakdowns after a year's hard use in the desert?

I went over in my mind some of the truck's troubles. For one thing, the radiator had long been in bad shape. In early May Bodil, Stig, and I had been returning from Colomb Béchar when the radiator started leaking badly. At Igli the lieutenant helped us with an old and honored desert trick, putting some handfuls of dates in with the water; fibers and pieces of skin from the dates would plug the leaks. Indeed, we were able to continue on our way; however, the effect didn't last, and although we repeatedly poured in more water, the engine boiled every few minutes, and we were forced to stop.

At that point we were twenty-eight kilometers from Béni Abbès. It was dark, and we had to spend the night in the truck. I thought that if someone came past, I would send a message to Béni Abbès, but there wasn't much traffic in the Sahara in summer; we hadn't seen one vehicle since we left Colomb Béchar.

At five in the morning I heard engine noise and saw the lights of a truck coming from the direction of Igli. It was the weekly bus on its way south, but it was going directly to Adrar without taking the side trip to Béni Abbès. Stig wanted to take the bus to the road fork and go by foot the remaining eighteen kilometers to Béni Abbès, at best about a three-hour walk. I couldn't let him leave; I was responsible for his safety, and eighteen kilometers in the rough desert is a considerable distance.

In the end Stig made a quick decision and jumped on the bus as it was leaving. I was worried but could do nothing. True, it would stay cool until the sun came up, but then it would heat up rapidly. I knew too well what happens if a person follows a wrong track and gets lost in the desert. A single day without ample water would probably be fatal. At seven a truck came by on its way north, and I sent a message to the lieutenant at Igli. Then we waited again. It was getting hotter. At nine I heard a car, and Kast, the physicist at the research station, came in a Land Rover and towed us to Béni Abbès. Stig was safe, and I could stop worrying.

By summer our truck had more problems. A crack had spread from the cab roof to the rear, as if the vehicle were about to split open lengthwise. More serious were the problems with the engine, the drive train, and especially the springs. We would be heavily loaded when leaving Béni Abbès, and it would be a miracle if we were not stranded for days on a mountain road with a broken spring or axle. Towing would be terribly expensive and probably impossible on the mountain roads. Not believing in miracles, I decided to avoid a mountain breakdown by driving only to Colomb Béchar, where we could put the truck on a flatcar and ship it by train across the mountains.

I said goodbye to Kast, the physicist who always was willing when we needed his help. I paid the expected last official visit to the captain to thank him for assisting us with housing and laboratory space and finding helpers for us. I was sorry to leave Lagrib, who had served us so unfailingly; I knew that I would probably never see him again.

Stig and I left Béni Abbès on July 8 at five in the morning so we would arrive in Colomb Béchar before it got too hot. Mohammed came along, to help us, he said, and besides, he was eager for a trip to Colomb Béchar. Our last night in Béni Abbès had been relatively cool and pleasant, the truck ran well, and we had managed to leave one day earlier than we needed to reach the train on which I had reserved a flatcar.

At Igli, seventy kilometers from Béni Abbès, we stopped for a few minutes to say goodbye to the lieutenant who had helped us several times. After another 125 kilometers we were only 50 kilometers from Colomb Béchar and would be there in about an hour. Mohammed pulled out a basket with Arabic bread, boiled eggs, and a piece of mutton, which we shared as I drove. I stopped so we could drink Mohammed's tea and some water. When we walked to the rear of the truck where we kept water, we found one door hanging open! We had not stopped since Igli and had no idea when the door had sprung open. All the big boxes were there. My two small fennecs were unharmed in their box, but we had lost many smaller items that we had put in last. The worst loss was Stig's briefcase, containing his travel documents and passport.

We turned back, knowing that unless we soon found everything, we would be forced to return to Igli to get gas. The heat was increasing, and we had to repeat the worst part of the road. First we found the shattered pieces of the goula that I had put in at the last moment to take home as a souvenir. We continued and found more things and, finally, when we were almost back in Igli, Stig's briefcase.

By then it was around noon and 45°C (113°F). We drank immense quantities of water, even though we had been drinking each time we stopped to recover a lost item. We got gas at Igli, but were advised to wait until it got cooler before driving on because the tires could overheat and explode. We drank more water and lay down to sleep on a burning-hot cement floor. When we decided to continue at five, we discovered two broken leaves in a rear spring. There were no spare parts for our truck, so we had to go on.

I drove slowly in the burning heat and was grateful for each kilometer we covered, for that meant one less to go. Sixty kilometers from Colomb Béchar, we stopped at a military post, where we found two more leaves broken. The body now hung on the rear axle. When we reached Colomb Béchar at ten, our truck had five broken leaves. Because of the mishap with the door we had driven nearly 400 kilometers instead of slightly less than 250 to Colomb Béchar, and had been on the road for seventeen hours, except for the few hours of overheated midday rest.

Late in the afternoon on the next day, with the springs fixed, we drove the truck onto the flatcar I had reserved, and our immediate worries vanished. We said goodbye to Mohammed, sad to know that we were unlikely ever to see this wonderful man again.

In the evening we returned to the train station and found that there was only one old third-class passenger car, which had two wooden benches in each compartment. We chose a compartment and lay down to rest. Most other passengers who looked in found seats elsewhere, but one man mumbled something about a family, left, and returned with his veiled wife and a boy of about five. He loaded an abundance of packages and baskets into the compartment, rolled out a mattress on the floor, and ordered his wife to sit. He and his son sat, one on each bench.

After the train started, Stig and I lay down again, although we couldn't stretch out as comfortably as before. Then, to our surprise, a small child began to cry. A two-year-old girl was with the mother, who apparently had carried the child under her ample white garments. The woman sat on the floor, now and then resting her head against me where I lay on the bench. The situation felt strange. The man, who treated his wife with no sign of kindness, spoke excellent French and seemed educated. Later, when she was lying down, he covered her with her veil from her feet to the top of her head. Gradually the veil slipped away from her face, and in the moonlight I could see that she was pretty and very young.

At daybreak, about four in the morning, we made a two-hour stop in

Béni Ounif, which gave me time to check on the truck. I noticed that the blocks under three of the wheels had come loose, and fastened them carefully. Stig and I moved to an empty compartment, where we slept again, freezing even under our blankets as the train continued through the mountain range.

The train made several stops, and a number of men boarded. I didn't know whether they were Arabs or Berbers, but they did not exchange a single word or greeting with us. They may not have known much French, but in their daily lives they could hardly have avoided picking up a few words.

The compartment gradually filled up, and Stig and I were squeezed into the two corners at the window. A man settled next to me, dropping himself halfway onto my thigh without apology as if he had sat on a sack of dates.

These people conveyed a strong feeling of hostility. It was about a year before the Algerian war of liberation, and the men may well have been among the tribes that organized the resistance. That spring we had visited the khaïd of Guerzim, who had been on a trip to Morocco. When the captain in Béni Abbès heard about our visit, he expressed displeasure; he undoubtedly knew about the brewing unrest.

We arrived at Saïda at eleven and found rooms in the hotel where Dick Houpt and I had stayed when we had engine trouble on our way to Béni Abbès in the fall. The next morning we unloaded the truck and continued on our way. The following day we reached Algiers and began the long procedure of shipping our goods back to Durham. We unloaded our crates and footlockers at Worms warehouse, had them numbered and strapped with steel bands, gave lists of the contents to the superintendent, and left.

We had the good fortune to spend our one evening in Algiers with Professor Bernard, the zoologist I had met at the desert conference in London two years before. He and his wife took us to a restaurant where the food seemed utterly luxurious in contrast to the dry bread and tea with powdered milk we had lived on during the last weeks in Béni Abbès. As we sat with clean napkins and a white tablecloth, it finally sank in that we had left behind Béni Abbès and the hard work, that we were no longer trespassers in a society where we were intruders, that we were back where we had come from and where we belonged. We had accomplished what we had set out to do and were on our way home.

In the morning we started our long journey to Denmark, driving the old truck west toward Morocco. We crossed to Spain at Gibraltar and continued north through France and Germany to the Danish border, where the two fennecs escaped notice by the customs officials. We arrived at Stig's house in a Copenhagen suburb in the middle of the night, and after a shower and a cup of tea I drove on, tired but eager to see my family again.

CHAPTER 9

Turtle Tears and Saltwater Frogs

In DENMARK I joined Bodil and the children at the old vacation farmhouse. Bodil's sister Ellen was there with her children and her husband, Niels Gunder Knudsen, a noted animal sculptor. Niels was a huge man, strong as a bear, with a round head topped by thinning red hair.

Niels and I got along well. I always felt closer to sculpture than to painting, and I liked Niels' work. We also shared an interest in animals, fishing, and, more than anything else, tools and how to use them and make things. Niels made many of his own tools and knew all sorts of tricks for handling difficult materials and solving technical problems. He preferred working in ceramics, a material he handled with exceptional skill; but he also worked in wood, bronze, or stone as the artistic problems at hand might demand. Niels took an immediate interest in

145

the fennecs and made a sand-colored ceramic sculpture of the female, capturing her watchful personality.

Over the years Niels became a close friend, and I have acquired many of his sculptures. Several have a special meaning for me. His portrait bust of Bent as a child communicates the thoughtful, almost dreamy quality of Bent's boyhood personality. His sculpture of a kangaroo rat sitting on its hind legs reminds me of how the many years of desert research began with the study of this little rodent in Arizona. Nor could I resist buying his bronze cormorant because it was while working with cormorants that I discovered how marine birds survive at sea without fresh water.

At the end of the summer we returned to Duke and the myriad responsibilities and demands of academic life. I prepared lectures, taught student laboratories, and had numerous other academic duties. I also had to write reports to the agencies that had supported our research and analyze and write up the masses of raw data from Béni Abbès. At Béni Abbès I had maintained the main notebooks and recorded the daily progress, but Bodil knew the material almost as well as I did. She made many of the calculations and graphed the results. We worked hard and made good progress, preparing several papers for publication.

Bodil's salary came from a grant awarded by the Office of Naval Research that required only that she continue doing productive research. With no teaching responsibilities, she was free to resume her studies of the kidney function of kangaroo rats while I took care of a mounting amount of other work—reports, correspondence, accounting, and whatever else piled up on my desk. Faced by demands that often seemed insurmountable, I became increasingly ineffective. Although I met grant application deadlines and gave reasonably well-prepared lectures, I tended to postpone dealing with problems. If I took a letter from the stacks on my desk that required a careful response and there was no urgent need to answer it, I stuck it under the pile and looked for something easier. My desk was filled with unfinished work, yet I often would just stare at the heaps of paper.

I recognized the pattern, and my inability to work effectively made me increasingly unhappy. Caught in a vicious cycle, I dwelled on my shortcomings and unfortunate mistakes in the past, brooding over how poorly I had handled my life. I compared myself with Bodil, who seemed well-liked and hard-working, successful in her research and in control of her life. I knew that others thought of me as successful; how could they know what a wretched creature I really was?

At home I wanted to be kind to the children, but often I was impatient and unreasonable. I demanded good behavior at the table, shouting instead of talking quietly to them. Frequently I felt sudden anger over trivial matters and disliked myself for it, yet I wanted to spend evenings with the children and would often read to one of them while Bodil took care of the others. Because Astrid was older, we had more time for her after we had said goodnight to Mimi and Bent.

I contemplated asking advice of a good friend, the psychiatrist Hans Lowenbach. Hans had lived in Norway for several years, and I knew about him from my time in Copenhagen because he had helped August Krogh by obtaining samples of whale urine from Norwegian whalers. I found my thoughts always coming back to a chance remark Hans had made. During an interesting conversation about other matters, he had casually mentioned that psychiatric intervention seems to have little enduring effect on people after the age of forty. This notion frightened me: I was thirty-nine, I was desperate, and I had to do something about myself.

One Sunday afternoon when Hans and I went for a walk in the forest, I told him how troubled I was about my work. He listened only briefly. He clearly didn't want me to become his patient; he wanted us to remain friends. Hans said he would arrange an appointment for me with Bernard Bressler, a psychoanalyst who was about to join the Duke staff.

In January 1955 I had my first meeting with Bressler. I poured out my concerns, told him how worthless I was, how little I accomplished, how often I got angry over nothing, what awful things I had done or contemplated doing, violent thoughts that frightened me, fantasies about suicide—a whole hour of baring my shame and fear, matters I had thought I would never tell anybody.

At my next appointment Dr. Bressler said he would accept me in analysis and laid down explicit conditions. Forty-five minute sessions every day, four days a week. No cancellation of appointments. No reading of books about analysis. No taking notes about the analysis.

Payment of all fees on time. I later understood that patients sometimes play games with their analyst by withholding payments.

I had an intense wish to change and subjected myself unconditionally to his requirements, arriving faithfully on time for every appointment, working hard to remember my dreams, and struggling to report all thoughts that came to my mind. I felt that if I meticulously followed Bressler's instructions, I would succeed, even though I had no clear concept of what I wanted to achieve. Although at times the sessions were agonizingly difficult to continue, I felt a great relief that I was doing something about my lifelong problems.

With one exception, I declined all invitations to speak at meetings I would have loved to attend. Bodil went alone to the yearly meetings of the American Physiological Society, which I had attended every year until then except for the year in the Sahara.

The relationship between a patient and an analyst has no counterpart in other human relationships. It has been compared to that of a confessor and a priest, but the similarity is superficial. The analyst doesn't sit in judgment about good and bad, about "sin" and punishment. The remarkable result of a lengthy analysis is, I am certain, an increased understanding of an individual's own unrecognized desires and motivations, and the development of a more honest human being.

I had expected to delve into childhood events and reveal the causes of much unhappiness. This was not what I found; no hidden memories or forgotten traumas suddenly came to the surface to resolve all problems, past and future.

I cannot put into words exactly what I achieved. In long and laborious sessions I brought up memories of stupid things I had said and done, deplorable words and acts I wanted to leave behind. Today I know I am the same person I always was, but I am also happier, more productive, and less depressed, and I spend less time punishing myself for past mistakes, imagined or real. In short, I can accept more of what has come my way—the opportunities, the successes, and a great deal of love and happiness.

At the beginning, Dr. Bressler allowed one exception to my regular routine. He permitted me to travel to a symposium in Socorro, New Mexico, on arid land development that I had agreed to attend. I had been invited to describe the camel research and its implications for grazing in semi-arid areas. In fact camels are far better suited than goats and sheep for animal husbandry in desert areas. Unlike sheep and especially goats, which chew every piece of vegetation to the roots and denude the areas around oases, camels take only a few bites from a shrub or bush and then move on to another. In addition, a camel can go for perhaps a week without drinking, whereas a sheep must be watered every day. Other conditions being equal, a camel can therefore graze over an area seven times as far from a well, covering an area forty-nine times greater (the square of the radius) before returning to drink.

At the conference I also met scientists from all over the world who had interests similar to mine; several became friends and invited me to visit them in Australia and Israel.

Bodil was now clearly established as a scientist and well recognized for her work, which had brought her into contact with other prominent kidney physiologists with whom she had carried out several research projects. Her reputation was all the more impressive given that she had no university education and her most advanced degree until then was in dental science. In recognition of her work on the renal function of kangaroo rats and her publication record, the University of Copenhagen awarded her a doctor of philosophy degree in May 1955 without the usual formal disputation.

In the summer of 1955 Bodil traveled to Maine with the children to continue her studies in renal physiology while I remained in Durham for my regular appointments with Dr. Bressler. In August, Bressler's vacation month, I joined my family at Salisbury Cove, where we lived in a small cottage near the sea, only a few minutes' walk along the rocky shore to the laboratory. The children enjoyed the opportunity to swim in the sea and hike in the forest and mountains, a different world from the hot and sultry summers of Durham. No wonder Bodil liked Maine; the laboratory was excellent, and the children could freely enjoy their summer vacation.

During the preceding winter Mimi's left arm had a few times become momentarily stiffened, somewhat as if she had experienced a cramp; a minor degree of epilepsy, the doctors said. I was disturbed to see that she now had more frequent episodes affecting her arm, and it seemed that more of her body was involved. These brief episodes didn't cause serious problems, but Mimi definitely was not improving.

The following spring, in May 1956, Father died after years of increasing incapacitation by Parkinson's disease. I felt grateful that his long suffering had come to an end, but more intensely, I felt a deep sense of loss, deepened further by regrets over lost opportunities for contact and communication. It was so final—there was now no way to obtain answers to questions I had never asked, no opportunity to tell him about my own joys and sorrows. My sister, Astrid, to spare me a transatlantic journey merely to attend the ceremony, did not inform me of his death until after the funeral. In many ways it was a relief to escape the need for a hasty trip to Norway.

A year and a half after I entered analysis, Dr. Bressler gave me permission to be away for the summer, and I was able to resume fieldwork and pursue a topic that had long held my interest: whether or not marine birds drink sea water. Nearly twenty years before, in 1939, I had worked on this problem, but then the war had come, and after that all my efforts were concentrated on the physiology of desert animals. Now I finally had the opportunity to return to the marine birds.

For the first time since before our marriage, I didn't ask Bodil to collaborate on a major research project of mine. I knew she found the bird problem intriguing, but I wanted to work on my own and be independent of her. I felt that until then I had shared all my best work with her and had given her much of the credit for it, but Bodil had never suggested that I should share in her kidney work. Since 1949, when she returned from Denmark after her father's death, she had worked on the kidney function of kangaroo rats, studies that she planned, competently carried out, and published on her own.

For the study of marine birds I asked two postdoctoral collaborators to join me in Maine: Humio Osaki from Japan and a former classmate of mine from Denmark, Carl Barker Jörgensen. We caught some young

cormorants, and to find out what effect sea water has on salt excretion, I gave one of them a liberal amount by stomach tube and placed the bird in a carefully cleaned plastic container. Within a minute or two I made the fastest scientific discovery I ever made. I noticed that the bird, with a quick movement of the head, shook off droplets of fluid that appeared at the tip of its beak. I sampled the clear liquid with a micropipette; it gave a massive precipitate with silver nitrate, revealing a high concentration of chloride. We were astounded, but the result confirmed what I had suggested decades before—that if salts do not come out one end of the bird, they must come out the other.

The very salty secretion is produced by glands in the bird's head and drips from the tip of the beak. Thus, if the birds drink sea water, the excess salt is eliminated, leaving a net gain of free water. Whether marine birds in the wild actually drink sea water is a question that is difficult to answer. Nevertheless, much of their food has a salt content high enough to necessitate the elimination of excess salt by the glands we had discovered. For simplicity we decided to call them salt glands. Our discovery received a great deal of attention from physiologists as well as the popular press, for no such gland was known in any animal, and it solved a long-standing problem.

I continued these studies over the next two years, both at Duke and in Maine. All marine birds we examined—gulls, pelicans, petrels, eider ducks, and so on—use the same mechanism to excrete excess salt. I had a marvelous collaborator in a Swedish friend, the animal physiologist Ragnar Fänge, who described the detailed anatomy of the salt gland and refined our understanding of its function.

Fänge and I then turned our attention to other marine vertebrates. How, we wondered, do such marine reptiles as turtles, crocodiles, sea snakes, and lizards handle their salt problems with kidneys even less suited than bird kidneys for the elimination of excess salt? Some other mechanism for elimination was needed.

We began by studying several species of marine turtles, as well as a marine iguana from the Galápagos Islands. When given a salt load, all these animals excreted fluids with an extremely high salt content. In the iguana lizard the salty fluid came out from the nares, excreted by a gland that appears to be anatomically analogous to the salt gland of birds. Turtles, however, excreted salt from a totally different gland, located in the orbit of the eye, producing tears with a remarkably high salt content.

This unexpected finding reminded me of a line from *Alice in Wonderland,* which I had read to my children: "So they went up to the Mock Turtle, who looked at them with large eyes full of tears." Fänge and I decided to pair the Carroll quote with one from Kipling's *Just So Stories,* in which the Crocodile, to convince the Elephant Child, "wept crocodile tears to show it was quite true." We hadn't studied marine crocodiles, but we used the quote anyway. (Many years later an Australian student reported that marine crocodiles have salt-excreting glands located on the tongue.)

Our manuscript started with these two quotes. I sent it to the prestigious journal *Nature* and wondered how the editors might respond to the use of storybooks as references. I was pleased and a little proud when it was accepted without comment.

In 1956 I was again invited to an international conference on arid lands, this time in Australia. Duke gave me leave from my academic duties, and I used the opportunity to go around the world. I decided to travel via Sweden and Norway and to return home via the Pacific. I flew from New York to Stockholm to see my sister, Astrid, and the two of us then went by train to Trondheim to sort through Father's belongings. Mother had moved to Stockholm, and we found Father's house empty but with everything still in place—books, furniture, paintings, and other beautiful things. We sorted through everything, reliving many childhood memories. We moved many items to Stockholm, but most of the larger furniture had to be sold. Patrick wanted only a couple of paintings, and I shipped some things to Durham, including a grandfather clock that was a wedding gift to my great-grandparents in 1845 and now stands in my living room. The wooden case is decorated with a magnificent example of the Norwegian peasant art known as *rosemaling,* or rose painting.

In 1956 going around the world was a great adventure. From Stockholm I flew to Amsterdam to begin the long haul to Australia in a propeller-driven Constellation. The plane's four engines droned away hour after hour until we reached Rome, the first of several refueling stops on our exhausting three-day journey to Sydney.

The conference took place in Canberra, the Australian capital, where the participants stayed in a modern and well-designed student dormitory. It was October, which corresponds to April in northern latitudes. The heating had been turned off, and we were miserably cold, but the conference was excellent, and I learned a great deal about arid lands and their management.

After the conference the participants went on a long field trip to the outback in a plane that could land on short grass strips, a two-engine DC-3 known as a Dakota. At the dry mining areas near Broken Hill we were introduced to the use of drought-resistant eucalyptus trees for land reclamation. The area has minimal rainfall, and for a year the eucalyptus seedlings had to be watered once a week. During the second year they were weaned from watering, but they had already set deep roots, so most survived. This and other new knowledge made both the conference and the field trip wonderfully successful.

I returned to Durham to my usual teaching responsibilities and my continuing research on marine birds. I still saw Dr. Bressler four times a week and felt that I was making progress in sorting out my past problems. Overall, though, my relationship with Bodil wasn't good. I didn't like it when Bodil announced she had been invited by the president of the American Physiological Society, Alan Burton, to present the prestigious Bowditch Lecture at the society's fall meeting; she was to talk about the camel studies. The Bowditch Lecture was to be presented by an outstanding young physiologist and had been presented only once before, by John Pappenheimer of Harvard University.

Although I had been the leader in the camel work, I reasoned that my age had been the reason Dr. Burton hadn't invited me instead of Bodil. I was forty-two, and the official age limit was forty; Bodil was thirty-nine. Shortly thereafter, however, the secretary of the society, Milton O. Lee, told me, "You must understand that it was a misunderstanding that the invitation wasn't addressed to you."

"But I am over forty and cannot be invited," I protested.

"No," he explained. "We didn't think the age limit had to be taken that seriously. You were the organizer and scientific leader of the expedition, and Dr. Burton wanted to invite you. I asked him if he was certain which of the two first names was yours and which was your wife's, and he said he knew. However, when he wrote the letter he chose the wrong one. I am sorry he made that mistake, and I wanted you to know how highly we think of you."

I appreciated Lee's kindness in telling me that they had planned to invite me. I knew he did not intend to cause difficulties, but the fact that Bodil had received an honor intended for me did not improve our relationship.

In the summer of 1959 the whole family went to Denmark, and then to southern Italy for two weeks. I explained that this might be our last vacation as an entire family because Astrid would be going off to college that fall. The real reason, however, was that my relationship with Bodil was deteriorating, and I began to feel that a separation or divorce was inevitable.

The trip carries good memories for me. After visiting Rome, we continued to Naples, where we boarded a bus for Amalfi, a small fishing village on the coast farther south. We enjoyed the nature, the sea, and the sunny weather. Even the stress of getting to Amalfi was worth it. The last stretch of road from Positano goes over mountainous country and descends steeply to Amalfi. The hairpin curves were so narrow that the bus driver couldn't make the turns without stopping in the curve with the front wheels at the edge, then backing up a few meters and again turning the wheels as far as he could to complete the maneuver. At each curve, as the driver struggled with the bus, we looked straight down at the blue Mediterranean, hundreds of meters below, the bus moving within inches of the edge. With white knuckles I gripped the edge of the seat, doing my best to keep the bus from plunging over the cliff. I told myself that if the driver was no good at this game of life and death, he and the bus would long ago have ended up in the blue water. I was exhausted and relieved when the ride ended and we had our feet on solid ground again.

As I had expected, the two weeks in Italy were the last reasonably peaceful time we spent together as a family. After we returned to Denmark, I left the others and went to Sweden to join my friend Ragnar Fänge at the biological station at Kristineberg to continue our studies of marine birds.

I felt happy and free and made friends with people there. One was Lewis Wolpert, an engineer from South Africa who had turned to biology, left his country, and now lived in London. After working hours I often spent time in the company of a tall and charming Swedish girl who worked as a technician in one of the laboratories. Sometimes we played a set of badminton in the afternoon or in the light summer evening. I hadn't played badminton before but rapidly became reasonably good at it. I knew that she also enjoyed my company, but I was determined not to let our friendship develop into a love relationship. Still, I was sad when she left, having discovered—after years in an unsatisfactory relationship—the pleasure and fun I could have in the company of an attractive woman.

A FTER DISCOVERING salt glands in marine birds, I began a detailed study of the physiology of these interesting glands. We had found that all marine birds have salt glands, and that there are similar glands in marine reptiles, such as turtles and marine iguanas. We found that seals have no salt glands, but we hadn't expected any, since the kidneys of many mammals can readily excrete large quantities of salt.

But I was curious about amphibians, such as frogs and salamanders. Normally frogs do not live in the sea; the high permeability of their skin would cause them serious problems in sea water because their blood and body fluids contain less than 1 percent salt, whereas sea water contains about 3.5 percent. A frog in sea water, it seems, should soon resemble a pickled herring.

I had come across a few reports of frogs and toads that live in brackish water, and even a mention of certain frogs that swim in full-strength sea water. This sounded incredible. If frogs living in the sea really existed, their physiological mechanisms certainly deserved a careful study.

In the spring of 1960 I planned to search for saltwater frogs where they had been reported, on the tropical coasts of Southeast Asia. Pete Scholander told me that Malcolm Gordon, a physiologist at the University of California at Los Angeles, was planning a similar study. "Why don't you two work together?" he asked.

I was more than happy to collaborate with Malcolm, and we planned to work at the Oceanographic Institute at Nha Trang, a town north of Saigon in Viet Nam. At the beginning of the summer I flew from New York to Los Angeles, my first trip in a modern jet. From there, Malcolm, a student named Hamilton Kelly, and I took off for Viet Nam to search for saltwater frogs. After stops in Tokyo and Hong Kong, we arrived in Saigon.

Before our departure, Scholander had arranged that we would be officially affiliated with his home base in the United States, the Scripps Institution of Oceanography. On our arrival in Saigon I discovered the advantage of this affiliation with an institution that had a semiofficial standing and was aided by the U.S. diplomatic services wherever it operated expeditions. At the airport in Saigon we were met by a representative of the University of Saigon and a car from the U.S. embassy. I wondered how to explain our instruments and chemicals to custom inspectors who understood no English and little French, but thanks to our friends from the university we cleared customs with no more than a perfunctory inspection.

It was the rainy season, and Saigon was hot and humid, with frequent thunder showers and sudden downpours. But getting drenched didn't disturb us; we were never cold in the steaming heat and were wet whether it rained or not.

At that time Viet Nam was in the relatively peaceful period that followed the departure of French troops, before the arrival of U.S. military forces. In the main streets of Saigon the cars, the restaurants, and the sidewalk cafés were all French. Even my hotel bed was very French and uncomfortable. I immediately removed the huge cylindrical bolster found under the pillow in all French beds, convinced that its only purpose is to break the necks of unwary foreigners.

We were warned not to eat milk, butter, or cheese because of the danger of undulant fever, brucellosis, though imported French butter and canned milk were safe. Unpeeled fruits were forbidden, and we were told not to drink the water and under no circumstances to eat uncooked vegetables or salads. The restaurants had a habit of decorating every plate with a sprig of watercress, which we always pushed gently aside, for even our moderate knowledge of tropical diseases and parasites transmitted by unclean water made us cringe.

Before beginning our search for saltwater frogs at Nha Trang, we bought some chemicals we needed but were not allowed to carry on international flights. As in North Africa, I was again frustrated by the three- or four-hour period after lunch when all shops were closed. Two days later we were on our way to Nha Trang in one of those old and trusty workhorses, an American DC-3, flying low over the meandering rivers, the flooded fields, and the small huts on the plains. Farther north the country became hilly, and we flew over jungle-covered mountains, low enough to discern an occasional flowering tree amid the solid greenery.

Pete Scholander and Claude ZoBell, a well-known bacteriologist also from Scripps, had arrived in Nha Trang ahead of us and informed us about local conditions and customs, the water, and the food. Despite all the precautions, we soon had stomach upsets. My first episode was not as violent as Malcolm's, but it lasted longer. Of course, we didn't drink the water. Even so, ZoBell tested the tap water and with a big grin declared that the bacterial count was so high that the water supply must be hooked up directly to someone's intestine.

The institute's personnel told us that we could safely drink water from the institute's filtering plant; it was stored in a large ceramic container where it supposedly was sterile. Dr. ZoBell repeated his tests and reported that the filtered water had an even livelier microbial population than the tap water. Thereafter we treated all water we used for drinking and for brushing our teeth with water purification tablets. ZoBell's tests confirmed that the treated water was indeed sterile.

Malcolm and I searched for frogs, but the countryside was still dry for lack of rain, and our task was hopeless. Frogs were sold in the market, but Malcolm identified these as ordinary frogs. There was no use in staying in Nha Trang to search for frogs we weren't sure existed, so we decided to move to Thailand to continue our search. We packed our equipment and flew to Saigon. Malcolm's wife, Diane, had joined us; the two of them decided to travel via Cambodia to see the magnificent temple ruins at Angkor Wat, while I would fly to Bangkok with our equipment and start the search for frogs.

Before leaving Saigon I wanted to visit the Pasteur Institute, where the French physician Michel Barme was studying sea snakes and their venoms. My interest was not in venom but in the snakes' physiological adaptation to sea water. Sea snakes, which are closely related to cobras, are unique among reptiles in spending their entire life cycle in the sea; other sea-dwelling reptiles such as turtles come ashore to lay their eggs. Sea snakes are only a few feet long, highly venomous, and brightly colored, no doubt as a warning to potential enemies. From my previous studies I knew that marine turtles and iguanas have functioning salt glands; now I was eager to find out if sea snakes have similar glands.

Dr. Barme was interested and wanted to help me. But how do you handle a sea snake safely while injecting salt or making it swallow sea water? I suggested we put one into a long glass tube and leave the tail outside so that I could inject a saline solution. But any secretion of salt that occurred would be in the head region, which wasn't easy to observe inside the glass tube. Neither Barme nor I wanted to let the snake stick its head out of the tube, and after hours of negative results we gave up, to the doctor's great relief.

In the middle of these experiments Barme left to attend to patients bitten by dogs that might be rabid. We were in the wet season, and Barme told me that the prevalence of rabies in Saigon was shocking; ten or fifteen newly bitten patients came to the institute each day. About 50 percent of the offending dogs were rabid, so all patients were treated with antiserum. Depending on the location of the bite, a patient had to return between eighteen and twenty-three times for repeated injections.

I quickly estimated that if only ten new patients came every day and each needed twenty repeat injections, the institute would have 200 persons to treat on any one day. In the dry season the number of bites increased to 40 or 50 per day, which would easily lead to 1,000 patients being injected daily.

I had seen skinny and sick-looking dogs roaming the streets of Saigon, many covered with sores and pus. "Why aren't sick dogs eliminated?" I asked.

"They should be," said Barme, "but for religious reasons people object to killing animals, and dog catchers are scarce. Feeding poisoned meat to stray dogs is more effective and has been reasonably successful. As killing a dog is against the religion, a dog is offered two pieces of meat, one poisoned and the other not. If the dog chooses the poisoned piece and then dies, it was the dog's choice."

That evening I flew to Bangkok, leaving sea snakes and sick dogs behind.

In Bangkok I was struck by the obvious signs of prosperity, in contrast to Viet Nam, where I had become accustomed to barricades and soldiers, although I never got used to soldiers who understood neither English nor French and merely pointed a gun at my belly.

In the morning I went to Chulalongkorn University, hoping to find some biologists who could help me track down the alleged saltwater frogs. An instructor, Twesuk Piyakarnchana, who luckily went by his first name, told me he had seen frogs swim in the sea at the university's biological station at Bang Saen.

Captain Faughn, the captain of a Scripps research vessel, whom I had met in Saigon, stopped by the university. He suggested driving down to Bang Saen, about 100 kilometers away, to look for frogs. The road ran straight for kilometer after kilometer alongside a canal, but the road was narrow, so we had to drive slowly to avoid hitting other traffic. The leisurely pace allowed me to look at the fields and meandering rivers, the wooden houses, the people, and especially the water buffaloes. The huge animals seemed gentle and easily handled; often a small boy sat on one and poked it with a stick. From some of the houses a long pole jutted out over the canal, supporting a huge disk-shaped net that could be lowered into the water by a pulley arrangement and later retrieved, with dinner wiggling inside. We returned to Bangkok without seeing any frogs.

Early the next morning I walked from the hotel to the river. I bought my breakfast from a Chinese woman who was frying bananas over a charcoal fire on a street corner. She put several handfuls of fried bananas in a paper bag neatly pasted together from sheets of a news release from the U.S. Information Agency. I got into the habit of buying fried bananas from her; she certainly helped to get the U.S. news widely distributed.

I continued my walk through the hot, dusty city. The water in the many canals was black and putrid, too polluted to harbor any life. Although the growing season was almost over, there was an abundance of tropical fruit for sale in the native markets. Some of the fruits I knew well—pineapples, mangoes, oranges, limes, and bananas, including my favorite, small and thin-skinned bananas about the size of my thumb. There were others I had learned about only recently—mangosteen, litchi, rambutan, pomelo—and many I had never seen before.

Even though it was the rainy season, the sun beat down until I felt a part of the dirty, smelly city. When I arrived at Chulalongkorn University shortly after noon, I had walked for six solid hours and felt that I had seen much of the real life in Bangkok. I was happy to sit and drink the boiled water that an American scientist offered. "The city water is probably safe," he explained, "but the pressure varies and may drop to zero in a tall building. If a hose is connected to a faucet and the other end is in a bucket of dirty water, the whole thing may get sucked into the pipes." I decided not to try the tap water.

Twesuk and a Hawaiian zoologist, Tetsuo Matsui, a scientist on board the *Stranger* who worked at Chulalongkorn when the ship was in harbor, drove me to the biological station, where Twesuk asked about frogs. A local man took us to his house, set high on poles at the edge of a mangrove swamp. When we arrived, the tide was in, and the water almost washed across his floorboards; at low tide, when the water receded, he lived high above a dirty, stinking mud flat. The man told us to look for frogs in the evening when the tide was out. Unfortunately, Twesuk could not stay for the frog hunt.

The moon was full, and the difference between high and low tide was remarkable, 3.8 meters, or twelve feet, according to the tide tables. After dark, Tetsuo and I crawled in the mud under the houses. Shining our flashlights over the mud flats, we saw small points of light, the strong reflections from the frogs' eyes, telling us where to grab. Several small boys appeared, seemingly out of nowhere, and followed us around. They shouted and laughed and had great fun, crawling in the mud and helping out as best they could. They had no flashlights and couldn't "shine" the frogs the way Tet and I did, but they enjoyed our hunt.

One of the boys attached himself to me and followed me closely. Once he tapped me softly on the shoulder and pointed to a stone, indicating for me to shine my flashlight under it as he tilted it. Beneath the stone sat a huge scorpion with its tail lifted, ready to strike. I expected the boy to kill the scorpion with a rock, as a ten-year-old boy would probably do elsewhere. Instead he slowly lowered the stone, being careful not to harm the animal. I will never forget the gentle movements of this child whose culture was so different from mine. How thoughtlessly we Westerners kill what in our narrow minds we consider vermin.

After a couple of hours we were covered with mud and had about forty frogs in my canvas bag. There seemed to be no doubt that these frogs lived at the edge of the sea and swam in sea water when the tide was in and covered their habitat.

We returned to Chulalongkorn University with the frogs, and to test their tolerance, we put groups of five frogs in various salt solutions. These frogs survived for days in fresh water as well as in salt solutions up to about 3 percent, not unlike the concentration in coastal sea water (full-strength ocean water is 3.5 percent salt). This was many times what ordinary frogs would survive.

We had now confirmed the reports that our frogs were exceptionally tolerant to salt solutions, and when Malcolm arrived after adventurous delays at the Cambodian border, we worked to find out what physiological characteristics distinguish them from ordinary frogs. First of all Malcolm confirmed that our catch was crab-eating frogs, *Rana cancrivora,* and very appropriately named, for we found remnants of small crabs in their stomachs, and in one even some pieces of a scorpion. I wondered whether the frogs are insensitive to scorpion venom or are clever enough to grasp a scorpion before it strikes.

For several weeks we worked hard all day, and in the evenings we ate dinner together, usually at a Chinese restaurant. After dinner we went to the fruit market, where I had my first encounter with durian, a notorious tropical fruit that most Western travelers, if they ever see one, find so offensive that few have tasted it. Early in the twentieth century the distinguished plant collector David Fairchild brought plants and trees from all over the world to the United States, but he never imported durian. In one of his books he explained why: durian, he said, smells like a mixture of garlic and Limburger cheese. Pete Scholander's language was more colorful; when he spoke about durian, he likened its odor to that of an outhouse, "and very ripe!" he added.

Durian is large and oval, about the size of a football, yellowish green, and covered with spines like a vastly enlarged horse chestnut or buckeye. Twesuk told us that when the fruits ripen, durian trees are so valuable they are individually guarded against theft of the fruit, and people say that a guard who falls asleep under a tree may be killed by a heavy fruit falling on his head.

We paid three dollars for our first durian, nearly twice what our good Chinese dinner had cost each of us. We took the smelly fruit to our hotel and opened it. Inside we found two or three brown, sausage-shaped objects. I found Pete's unadorned description of the smell more apt than Mr. Fairchild's, for there was no doubt what these longish brown objects and their overwhelming odor suggested.

My usual curiosity about food won out, and I tasted one. At first, it was very sweet and pleasant, but as I tried to swallow, I felt reverse peristalsis in my throat. The look, smell, and soft, creamy consistency were overpowering. Even so, I managed to finish a whole segment. Malcolm's student Ham also ate one, but Malcolm and Diane refused. The next evening Ham and I agreed to try it again and went to the market to buy one. I was astonished; it seemed that both odor and taste had changed. The fruit now had a characteristic, fruity fragrance that I found immensely attractive. Ham and I were hooked on durian, and from then on we bought one every evening. During the day Ham talked about little else but our planned evening excursion to the fruit market. Fortunately, we lived in a simple hotel where few foreigners stayed. In the large tourist hotels, guests were prohibited from bringing durian into their rooms. Smuggling one in would be difficult because even a well-wrapped durian would leave a strong odoriferous trail.

Malcolm, Ham, and I went on additional hunts to catch more frogs, and we established beyond doubt that they tolerate high salt concentrations. Their blood and body fluids had a total osmotic concentration (the sum of salt concentrations and other dissolved substances such as urea) similar to that of the salty water we kept them in. However, only about half of the osmotic concentration was made up of salts; the remainder was primarily urea. These animals retained urea, a normal excretory product, in their blood to help match the osmotic concentration of the water without increasing salt concentrations to a level the frogs couldn't tolerate. This trick is the same one used by sharks and rays: they too retain urea until the osmotic concentration of the blood matches that of the sea water, although the salt concentration remains far below that in the surrounding water. Thus, two groups as disparate as frogs and sharks have arrived at the same solution for survival in sea water, a typical example of what zoologists call adaptive convergence.

In a few weeks we had the answers we were seeking, and Malcolm left for home. I continued to Australia with stops in Kuala Lumpur and Singapore. I was to join Pete Scholander on the Cape York Peninsula, where his research group was studying how mangroves manage to thrive with their roots in sea water.

THE PROFESSOR of zoology at the university in Kuala Lumpur was an American, John Hendrickson, who went out of his way to make my short visit unforgettable. He showed me the modern university, the city, and the local market and told me about fruits and foods that natives cooked and sold in the streets. He drove me around the countryside, through rubber plantations and past tin mines. On the way into the mountains we drove through thick stands of tall fern trees that reminded me of textbook illustrations of Carboniferous forests.

I asked John if durian was grown in Malaysia. "Yes," he replied, "but you can't buy any in Kuala Lumpur."

"But why?" I asked, astounded.

"It was banned by the British during colonial times because of its dreadful smell, and the ban on selling them within the city limits has remained."

"Have you learned to like durian?"

"No," he said; "neither I nor my wife has, but my children love it." I suggested that we try to find durian for the children, and on the way back to Kuala Lumpur we stopped at a modest roadside stand where the durian were smaller and cheaper than those I had bought in Bangkok but had the same wonderfully strong, fruity aroma. I bought seven of them. The smell lingered in John's car for days after we removed the fruits. I thought it was wonderful; John tolerated it without comment.

I resumed my journey to Australia and to Thursday Island, where several biologists from the University of California worked at a small biological station. A few days after I arrived, a member of Pete Scholander's team came in a small motorboat and took me to Pete's primitive camp on the Cape York Peninsula, near the mouth of the Jardine River, a three-hour ride.

Pete and two colleagues were interested in how mangroves adapt to growing with their roots in salt water, and with a group of Australian scientists he was also studying the physiology of the natives, who dived for pearl oysters without any technical equipment.

The camp was pitched on a small sandy spit only a few steps from the ocean; clean sandy beaches seemed to stretch endlessly in both directions. My first evening there, eleven huge female turtles came ashore along about half a mile of the beach to deposit their eggs. Each female digs a nest in the sand just above the high-tide mark. When she is digging and laying eggs, she is oblivious to humans and proceeds as if nothing could keep her from completing her obligations to the next generation.

She digs by forcing one strong hind flipper slowly, like a scoop, into the sand beneath her. Then, with a sudden jerk of the flipper, she throws a scoopful of sand far behind her. Then the other hind flipper goes through the same routine, a slow scoop and a sudden flip that throws the sand far away. She continues until the hole is about half a meter deep, remains immobile as her eggs drop into the pit one by one, then scoops sand back over the eggs and heads for the ocean at a slow, deliberate pace. I counted fifty-one eggs at a nest I observed from the time the turtle arrived until she left for the sea. In another nest the female dropped sixty-four eggs.

Months later the young hatch and dig up through the sand. Once out of the nest they head unfailingly to the water. The males never leave the water again; only the females set foot on land when, years later, as adults they return to repeat the cycle.

Turtle eggs look like Ping-Pong balls and have a soft, leathery shell. Most peculiarly, when I dropped an egg on a hard surface, it bounced a little. When we made scrambled eggs, the white wouldn't coagulate; the yolk, in contrast, became hard and gritty. The result was a grainy mass of cooked egg yolk with numerous lumps of uncooked, gelatinous white. However, the scrambled eggs were a welcome addition to the regular camp fare of coffee and condensed milk.

I stayed with Pete for a week, helping with analyses of the salt concentrations around the roots of the mangroves. On the way back, I remained for some time at the research station on Thursday Island, where there was more laboratory work to help with.

Before starting on my way home, I visited Australian colleagues in Brisbane, Adelaide, and Melbourne. I again saw Derek Denton, whom I had visited on my first trip to Australia several years before. His work fascinated me, and I admired his success in clarifying the accurate regulation of salt appetite in ruminants.

My first book, *Animal Physiology,* was published in 1960, the same year I traveled to Viet Nam, Thailand, and Australia. It was one in a series of twelve small volumes that together could serve as a college biology textbook. The market for introductory biology texts was crowded with heavy volumes; one book might begin with the structure of the cell, another with genetics, and a third with ecology, or perhaps with the classification of living organisms. Because the twelve small books were independent, teachers could combine several or all of them in the desired sequence to suit any particular freshman course. For example, one teacher might begin the course with ecology, whereas another might not include it at all because the students would later meet this subject as a required course.

The invitation to write this book came from the biochemist William McElroy, then a professor at Johns Hopkins University and later director of the National Science Foundation. He wanted a volume about physiology for the series he and the botanist Carl Swanson had planned. The series of small biology books represented a new approach that was soon copied by other publishers. Mine sold well, went through two revisions, and was translated into more than a dozen languages.

During the summer of 1961 I wrote a more important book, *Desert Animals: Physiological Problems of Heat and Water.* For years I had wanted to put together what was known about the responses of desert animals to heat and lack of water. By 1961 I had an extensive collection of reprints and notes. I had thought a great deal about the book and had a clear concept of how to organize it.

That summer, when Bodil went to Maine with the children, I stayed in Durham and worked at home, away from telephones and other interruptions, from early morning until I went to sleep, stopping only briefly for meals. I had already sorted my materials for the sixteen chapters I had planned. When I started writing a particular chapter, I spread all the materials on the bed and on the floor in the bedroom. I would lie on my stomach, alternately reading and dictating into a tape recorder as I worked through books and notes. On Monday and Tuesday of each week I dictated a rough draft for one chapter, based on reprints and articles I had read over the weekend. On Wednesday my secretary, Anne Mincey, transcribed my dictation while I read the typed copy of the previous week's work. On Thursday and Friday I read and corrected the transcript of Monday's and Tuesday's dictation and gave it back to Anne. On Saturday I started reading for the next chapter, to be dictated the following Monday and Tuesday, repeating the intensive routine. Meanwhile, Anne typed a clean copy of the drafts I had worked on the preceding Thursday and Friday. She was exceedingly accurate and fast; best of all, her patience was unlimited.

After fifteen weeks, only one chapter remained to be written. In September of the following year I completed the last chapter and sent the manuscript to the publisher.

The book covered almost everything that had then been written about the physiology of desert animals. To avoid the disparate styles of graphs and illustrations from the many papers quoted, I redrew most of them in a uniform style.

The book was well received. One reviewer, Tom Cade, then at Syracuse University, praised it for its many suggestions for further research. Other colleagues wrote to me to say that it was both easy and fun to read and would be useful in their own work.

But the most rewarding comment came years later from Torkel Weis-Fogh at the University of Cambridge. He used the book as required reading for every incoming biology student because, he said, he wanted all beginning biologists to know that good science can be interesting, fun, and a pleasure to read.

I was pleased that a specialized scientific monograph was even translated into Russian. It was reprinted twice, and in 1979, when it had been out of print for several years, Dover Publications in New York reprinted it once again.

Writing the book on desert animals brought my thoughts back to camels and important questions we had not yet answered. For example, we had never obtained reliable measurements of the metabolic rates of camels or of evaporation from their respiratory tract. When I had tried to measure the metabolic rate of camels in Béni Abbès, I had found it so low that I doubted the accuracy of my equipment; and I had been unable to measure respiratory evaporation because we had no suitable equipment. We knew that in kangaroo rats the respiratory evaporation is much lower than we initially could explain; what might it be in camels? To understand their water balance fully, we would have to find out how much water they lose by various routes.

During the fall of 1961 I organized an expedition, this time not to Africa but to central Australia, where camels had been introduced in the nineteenth century to serve as pack animals. Over the years these animals had gone wild and formed a thriving feral population. Australia offered several advantages for us. Instead of working under the primitive conditions of the central Sahara, we would have research facilities and excellent technical help readily available.

I wanted to work in the hottest and most arid region and obtained permission to work at a research station near Alice Springs, a small town almost at the geographic center of the continent.

My research group consisted of my gifted student Gene Crawford, now a professor at the University of Kentucky, and my colleagues Ted Hammel of Yale University and Ken Rawson of Swarthmore College. In early January 1962 we arrived at Alice Springs to work during the heat of the Australian summer. We were met there by Alan Newsome, who was attached to the agricultural research station at Alice Springs and worked with us while we were there. Not only did he have three wild camels captured and waiting for us; he had also hired two aborigines to care for our animals. To weigh the camels, Alan and a mechanic welded a tall, strong A-shaped scaffolding of steel pipes.

We had brought the laboratory equipment we needed, and in a couple of months we completed accurate measurements of both metabolic rate and water loss in several camels. To measure the respiratory water loss, we equipped a camel with a mask that permitted us to collect all expired air; this air was pulled through a canister that absorbed water, and the increased weight of the canister gave us a direct measure of the amount of water the camel had exhaled.

The water loss was so low that it could not be explained by any known physiological mechanism, yet our equipment was in perfect order and carefully calibrated, so there was no doubt the measurements were correct. The question was, by what unknown mechanism were camels able to retain water so efficiently?

As anticipated, we found that camels have lower metabolic rates than usual for animals that size. Also as we suspected, the camel's metabolic rate increases with increasing body temperature, and to calculate heat balance, it was important to know by exactly how much. We knew from Béni Abbès that the camel's body temperature in summer may rise to 41°C (106°F) without harmful effects; now we knew that temperature has a considerable effect on metabolic rate.

Metabolic rate also varied according to the camel's degree of dehydration. If the animal was deprived of drinking water, its metabolic rate decreased. Because both temperature and dehydration affected metabolic rate, and body temperature in turn was affected by the degree of dehydration, it was difficult to separate the effects of the two factors.

On our return to Duke, Gene Crawford analyzed all the data, using a slow, iterative process. When, after several months, he finished his careful work, I realized that a computer analysis using multiple regression analysis would have been much quicker and easier, but now we could at least use it to check the accuracy of Gene's calculations.

At that time computers were not yet in common use, and most biologists had not even seen one. The Mathematics Department at Duke told us how to put our data on punched cards and run them through the computer for a job that a pocket calculator today does in a few moments. To our relief, there was complete agreement between Gene's analysis and that of the computer.

Gene and I were rather proud of our computer analysis, and we were amazed when our manuscript was rejected. It was criticized by a referee who wrote something like "This manuscript is a classic illustration of the failure of a powerful tool (a high-priced calculator)." I wondered how much our referee knew about computers. I replied to the editor that the iteration by hand had taken a couple of months, whereas the computer, after the cards were punched, took a few minutes and cost between six and eight dollars, and we therefore didn't understand what "failure" the referee had in mind. The manuscript was accepted.

On the way home from Australia I stopped over in India. In Calcutta I was horrified by the poverty and by the sick and dying people in the streets. Cows walked around freely, and I was told that any driver daring to honk at a cow might be pulled out of the car and killed, or the car doused with gasoline and set afire with the driver inside. At a greengrocer's stand I saw a cow eating undisturbed from the displayed vegetables; no one would dare think of chasing it away. In another street a cow was lying next to a parked Rolls-Royce, peacefully chewing her cud. If the owner of the car wanted to leave, he would have to wait patiently until the cow moved.

In Benares I stayed at Clark's Hotel, an establishment dating from colonial times where tea is served in the garden each afternoon. The curry I had for dinner was the best I had ever tasted, so I asked the head waiter if I could buy some of their curry powder to take home with me. He explained in his modest English that the cook bought the curry in the market each day. Yes, this I understood, I told him, and I wanted to buy some, for I like curry, and this particular curry was better than any I had tasted. "But," the man explained, "the cook doesn't use a curry powder; he buys the separate ingredients." Yes, this I also understood, and I wanted to buy some of the separate spices. After more explanations, I finally understood why the waiter couldn't help me take curry home: many of the ingredients were fresh, green herbs; only a few were dry spices.

In New Delhi I was met at the airport by the UNESCO representative, Mr. Swarbrick, whom I had met when his organization sponsored our camel research in the Sahara. He invited me to his home for dinner, and the next day his twenty-year-old son took me on a tour of the city. I saw the famous Iron Pillar, erected in the fourth century. It is over twenty-four feet tall and sixteen inches in diameter; it is difficult to comprehend how so much iron could have been forged into one piece so long ago; another mystery is why the surface of the pillar, after centuries of exposure, has no trace of rust. What secrets did metallurgists possess 1,600 years ago?

Then I flew to Jaipur, and from there on to Jodhpur, to visit the deserts to the west and to meet colleagues at the Desert Research Institute. The camels in this area were the biggest I had ever seen. I also saw an elephant, decorated and painted with intricate patterns, at the head of a procession of colorfully dressed people participating in some celebration, perhaps a wedding. I asked a man the cost of an elephant;

about 10,000 rupees, or $2,000, he replied. Cheaper than a car at home, I thought; my last Ford had cost nearly $3,000. Wouldn't it be impressive to come to work on an elephant? But feeding it would cost perhaps 40 rupees per day. Gasoline was less expensive, and the car ate nothing when I didn't use it. I decided not to change my mode of transportation.

In one of the streets of Jodhpur a yogi wanted to tell my fortune, and I gave in to his urgent pleas when he offered to do it at half price, only five rupees, or about a dollar. Back at the hotel I made careful notes of what he had told me in his broken English so that I could check his predictions. "Will die at eighty-four years, die quick, like heart failure, no trouble, no sick, no harm. Have had money trouble, no more. Domestic trouble. One time hospital. Female, name letter M good for you. May, lucky month, good news, respectable life, high position."

The yogi showed an impressive ability to read the life of a person he had never seen before. Most of his predictions were correct—or so they seemed until I reflected further. I hadn't had money problems since I was a student. Domestic trouble certainly was no exaggeration, but hasn't almost everybody at some time had some domestic trouble? I had been in the hospital for my appendix, but there had been a couple of other times too. As for the letter M, any male whose wife's or sweetheart's name doesn't begin with M would probably think of his mother and consider that guess correct. For me, it wasn't until much later that the letter M came to signify the happiest years of my life. As for the eighty-four years, I still don't know, but most people would be pleased to know they have a long life ahead of them, and dying suddenly instead of wasting away with cancer would appeal to anyone. About a respectable life I must let others judge.

From India I went to Israel to meet Amiram Shkolnik and other scientists at Tel Aviv University whose studies of native desert animals and plants keenly interested me. I was welcomed most cordially and invited to accompany them on a field trip to the Israeli deserts.

Five of us went in a jeep to the Dead Sea. Instead of taking the road to the south end, we followed a trail eastward on the rocky high plateau until we reached the escarpment at about 600 meters above sea level. The driver stopped, and we all got out. Here, the land fell off nearly vertically and disappeared in an ocean of haze. We looked down, unable to discern the water 1,000 meters down, at 397 meters below sea level. I was still feverish from a continuing battle with intestinal microbes picked up in India, and I felt sick to my stomach as I saw two vultures, black and foreboding, soaring in wide circles as if expecting us to be their next meal.

I had no desire to ride over the edge and plunge into the abyss with nothing but bottomless fog. The driver, Amiram Shkolnik, said that it was too dangerous to ride with him; the jeep might overturn and tumble down. We looked on as he drove over the edge; then we followed on foot down the steep, winding trail, carefully picking our way over boulders and loose rocks that slid away under our feet.

At the bottom Amiram told me that the trail had been constructed by the army as a secret access to the Dead Sea. It was not intended for any other use. On the way down he had kept the jeep in four-wheel drive and in the lowest gear. He hadn't used the brakes, for if the wheels locked, the jeep would slide out of control, and he would have had no way to stop.

I was rewarded with a swim in the Dead Sea, where the salt content is about ten times higher than in sea water. No animal life is possible; over 300 grams of salt is contained in a liter of the concentrated brine, as opposed to 35 grams in the ocean. It is so salty that a person floats high in the water, and swimming is nearly impossible.

What I most enjoyed in Israel was the intellectual enthusiasm I saw in every scientist I met. They all wanted to show me the wonderful scientific opportunities virtually at their front door. I was pleased and satisfied, though exhausted, when I left, certain that I would be back.

On my way home I stopped briefly in Stockholm to see friends and family, then spent a couple of days in London with Lewis Wolpert, my friend from the summer at the biological station on the coast of Sweden. One day he arranged a theater evening for us, to include his wife and their friend Agi Bene, a psychoanalyst who later became my good friend.

From London, I began the last leg of my homeward journey, looking forward to seeing the children. I had been away for a quarter of a year, a long time when children are growing up.

Part III

Changes

[1962–1975]

Living Alone

By September 1962 Bodil and I had agreed that separation and divorce were inevitable. I moved into an apartment not far from Duke, taking my books, my bed, the old desk I had inherited from Grandfather, and some paintings and the grandfather clock from my parents' home in Norway. Bodil remained in our house with the children. Astrid, who was almost twenty-one, was in her last year at Oberlin College. Bent was eighteen and Mimi nearly fifteen; both continued their usual schooling and other activities, so superficially the change wasn't overwhelming for them.

I was relieved to be on my own and to arrange my life and apartment as I wanted. Finally I was free from the constant tension that for years had overshadowed my daily life. I discovered, however, that a single person doesn't fit in well among the couples that make up a university community and most other social groups. It would be awkward to invite old friends to a party in my small apartment, so I didn't. Although some good friends continued to invite me for dinner or to a picnic, I was alone more than I was used to.

One steadfast friend, Ed Horn, who was an assistant professor when I first came to Duke, had advanced to become chairman of our department and a very effective leader. I liked his straightforward personality and his no-nonsense manner. He was short and squarely built, his close-cropped hair was dark and turning gray, and with his strong, deep voice he might seem brusque or even unfriendly to those who did not know him.

When I was living alone, Ed often invited me over for a weekend dinner or to spend a Sunday afternoon with him and his family. Frequently we discussed new appointments and possibilities for improving the department.

I continued my work, spending long hours in my office and the laboratory. I ate at home and almost never went to restaurants or to the university dining rooms. I felt uncomfortable eating alone in public, sitting by myself with nobody to talk to, and reading at a restaurant table didn't feel right, except perhaps at breakfast. I could also eat better at home, for I was used to cooking and could bake my own bread. Still, spending a great deal of time on cooking only to eat alone didn't appeal to me, so I seldom prepared elaborate meals.

I remembered the Jewish refugee from my student days in Copenhagen, Mr. Knappeis, who cooked his modest meals, set a small table, served the food, and left his room, whereupon he reentered to sit down to a civilized meal. Like Mr. Knappeis I sat at the table to eat instead of grabbing something on the run. But my meals became irregular and often late, for there was always work to do, and why go home to my own company?

At least I was moving ahead with my work and excited about my research, in contrast to my first years at Duke, when I didn't seem to accomplish much. But I made little effort to meet new friends, and I spent too much time at work. Besides, I knew nothing about how single persons find new acquaintances. I didn't think of evening classes, political meetings, or other such opportunities, and, being totally indifferent to religious enterprises, I never considered church activities.

In the fall of 1962 I wrote the final chapter of *Desert Animals*, to be published by Oxford University Press. I asked the press to print a dedication: "To the Memory of My Father." This request signified a profound change in my feelings. Although as a child I had feared my father, I had slowly come to realize that he was a very unusual and admirable person. I thought how pleased and proud he would have been had he lived to see the results of my work. He was a good scientist and a good biologist and had a keen appreciation of quality. During the years I spent in analysis I had slowly come to view him as very special.

I stayed busy with daily work and teaching, and I received many invitations to lecture at other universities and at scientific meetings. Over the years, the requests became more numerous.

In the fall of 1962 I was invited to lecture to the Harvey Society in New York, a prestigious group of eminent medical scientists in the area. I was free to select the topic and found myself torn. Should I talk about desert animals or about marine birds and their water problems? As I was contemplating my options, it dawned on me that my work had a unifying theme: nearly all my studies involved questions of how animals adapt to hostile environments and how they solve problems of securing water. Not only desert animals, but also marine birds and saltwater frogs fitted in, each having evolved a different solution to what is basically the same problem.

At the Harvey lecture, my audience included many famous scientists, and in my nervousness I included too much information. I talked about sea gulls and crab-eating frogs, about camels and kangaroo rats, speaking too fast, trying to cover everything. But the audience evidently enjoyed the lecture, and afterward I was elected an honorary member of the society.

In December that year I lectured at an international UNESCO conference in India. The topic was animal life in deserts, an important subject for the arid areas in Rajastan and for many other regions in the world. I had become the world authority on camel physiology and on desert animals in general and was treated with great respect.

On the way home I spent a few days in London, where I saw Agi Bene, the psychoanalyst Lewis Wolpert had introduced me to the year before. She was a warm and engaging woman, and I felt comfortable in her company. We had corresponded since Lewis introduced us, and now we rapidly became close friends. She understood my continuing battles with unresolved problems, and I valued her nonjudgmental acceptance of my past struggles. She told me about gruesome experiences in war-torn Budapest and the atrocities when Russian troops invaded the city. Later, whenever I was in Europe, I stopped over in London to see Agi. Our friendship was important to both of us; I enjoyed her interest in literature and music, and we went to museums and on long walks on Hampstead Heath, that small piece of London that is as close to nature as is possible in a giant metropolis.

In December 1962 I received a letter from the president of Duke University, Deryl Hart, informing me that the board of trustees had approved my appointment to a distinguished professorship with the title James B. Duke Professor of Zoology. I was pleased, but I much preferred to be recognized as a physiologist and asked if my title could be changed to Professor of Physiology, as it had been when I came to Duke in 1952. My request was readily granted. Years later I also received an appointment to the Physiology Department in the School of Medicine.

In May 1963 I received a telegram from a small group of distinguished physiologists congratulating me on my election to the National Academy of Sciences. This marvelous surprise boosted my often flagging spirits. It meant that Academy members considered me among the most distinguished scientists in the nation and worthy of election. I was less than fifty, younger than most biologists when they are elected.

I immediately went to tell Ed Horn about the wonderful news and found him in his office, puffing away at a cigarette as he usually did while working. In his matter-of-fact way he congratulated me. At that time, few scientists at Duke had been elected to the National Academy, and he knew that the honor reflected favorably on his department as well as the university. The local newspaper carried a small note about the election, but there was no special recognition from the university and no change in my usual life and activities.

In early June 1963 I flew to Oberlin to attend Astrid's graduation. Bodil was not there, and I didn't feel I could ask Astrid how she felt about her mother's absence; for me it was a relief. I enjoyed the visit and was impressed by the high quality of this small school. Teaching was obviously its most important function. In many ways Oberlin reminded me of Swarthmore, where Astrid had started her schooling in kindergarten.

During graduation I visited an exhibit of student work at Oberlin's fine art museum. Among the many paintings there I found myself returning several times to a large impressionistic painting of an ocean of flowers in hues of red and orange and yellow, loosely arranged around a green ceramic bowl. Finally I asked an official-looking person if it was for sale. The sensible reply was a suggestion that I ask the student artist. The painting was unsigned, but I learned that one of Astrid's classmates, Linda Merritt, had done it. Linda was willing to part with it and it now hangs on the wall near the dining table, giving me pleasure each time I sit down for a meal.

When I returned to Durham I received an invitation from the University of California at Davis to serve as a Regents Lecturer during the fall term of 1963. Ed Horn immediately encouraged me to accept. Rather than pointing to my duties at Duke, he said it would do me good to be away for a while. He understood that in spite of recognition and successes, I was still grappling with doubts about myself. I was grateful for Ed's support.

On my way to Davis I stopped in Washington to attend the centennial meeting of the National Academy of Sciences, my first such meeting as an elected member. President John F. Kennedy addressed the body, and hearing his speech was unforgettable. I was already one of his admirers, and it was exciting to see him in person and to hear his forceful expression of optimism.

From Washington I flew to Albuquerque, New Mexico, to meet Agi. She came from Topeka, Kansas, where she was working at the Menninger Foundation, perhaps the world's greatest center for psychiatric training and research. In a rented car we took a leisurely trip to San Francisco through the spectacular desert areas of Arizona and Utah that I had come to love.

From San Francisco Agi flew back to Kansas, and I continued on to Davis to give my lectures. I was received as an honored guest and was invited to a great many official receptions and cocktail parties. Faculty members and the chancellor of the university, Dr. Emil Mrak, also invited me to lunches and dinners. As an official guest I lived in an elegant guest apartment and had a university car to drive.

The second day I was at Davis, my faculty host, the biochemist Art Black, asked for my help in filling out the employment form needed for payment of my honorarium. He asked for the usual information—date and place of birth, address, and other details. Then he turned the employment form over for me to sign. Printed above the signature line was a loyalty oath: "I do solemnly swear (or affirm) that I will support and defend the Constitution of the United States and the Constitution of the State of California against all enemies, foreign and domestic," and so on.

I explained to Dr. Black that I was certainly willing to support the Constitution of the United States, but as a resident of North Carolina how could I be required to defend California's constitution against all enemies, and perhaps bear arms in its defense?

"What does the constitution of California say?" I asked. Black admitted he had never read the constitution, although he had signed the loyalty oath. He doubted that any of his colleagues had seen or read it, and he fully understood how ludicrous it was to require a visiting lecturer to defend it.

Before I would reconsider my refusal to sign, I wanted to see a copy of the constitution. The university library rushed one over. I expected a thin formal document; in my hands was a medium-sized book. I leafed through some pages and came across material pertaining to the legislature and the court system, matters appropriate for a constitution. But I also found regulations for harbors and schools, for water rights and cemeteries, even for pawnshops—matters I had no desire to defend on behalf of a state in which I would spend no more than a few weeks.

While at Davis I asked many members of the faculty if they had read the constitution, but no one had. They were amazed when I told them some of what it contained. Nevertheless, they all had signed the oath lest they be fired as subversives.

I retained as a souvenir a copy of the form I refused to sign, and then I gave the lectures, fulfilling my part of the agreement with the University of California. The university, however, never paid me for the successful lecture series I enjoyed giving while the chancellor and faculty continued to treat me as a welcome guest.

On November 22, one of my last days at Davis, I returned to my apartment shortly after lunch to pick up some slides. I sat down at the table and turned on the radio; then I heard the horrifying news that President Kennedy had been shot by a sniper in Dallas, Texas. Shortly afterward the announcer informed the country that the president had

*The author's parents, Sigval Schmidt-Nielsen and
Signe Sturzen-Becker, at their engagement party in
Stockholm in 1906.*

*The house in Trondheim, Norway,
where the author grew up.*

The author with his oldest daughter, Astrid, when she was four years old (1945).

FROM THE LEFT: *The author's wife, Bodil; her father, August Krogh; and the author in Arizona in 1948.*

The author with his family and field vehicle before setting out for the Sahara Desert in 1953.
FROM THE LEFT: *Bent, Astrid, Mimi, the author, and Bodil.*

The author's three children on board MS Foria *on the way to North Africa in 1953.*
FROM THE LEFT: *M. Cremadés (cook on board* Foria*), Mimi, Bent, and Astrid.*

Research team with camels in Béni Abbès in 1954. FROM THE LEFT: *Bodil Schmidt-Nielsen, Dick Houpt (with donkey), Mohammed ben Fredj, the author, and Stig Jarnum.*

The author measuring the respiration of a cooperative camel (1954).

The author hoisting up a camel for weighing.
TO THE LEFT: *Mohammed ben Fredj.*

The author with the desert fennec that grew up in his pocket.

*Professor Richard Taylor measuring
the body temperature of a desert snail
in the sun (1969).*

*Professor Torkel Weis-Fogh admiring
a huge jelly mushroom in Duke
Forest (1968).*

*The author with Pete Scholander,
looking for surfacing dolphins
(1964).*

*Margareta
Claesson, the
author's wife,
in their home
(1985).*

The president of France, Valérie Giscard d'Estaing (FAR LEFT), greeting the author at a reception at the Palais de l'Elysée (April 2, 1979).

The author accepting the International Prize for Biology, with Emperor Akihito and Empress Michiko in attendance (Tokyo, November 30, 1992).

died. I was profoundly touched by the tragedy, a disaster for the country. Only weeks before, I had seen the president and heard his dynamic speech to the Academy members; I put my head in my hands and cried.

Back in Durham I resumed my routine, often working late into the evening, for my apartment was useful for little besides eating and sleeping. I had lived in single-family houses most of my life, and apartment living had no appeal; I much preferred a piece of land to put my feet on.

Astrid now lived in Washington, but she often visited Durham, and I told her about my wish to build a new house, although it seemed too extravagant for a single person. "Why shouldn't you?" she asked. "You can afford it, and you don't like apartment living." Her matter-of-fact words helped, and I couldn't think of valid counterarguments. I wanted to leave the apartment, but I also wanted to remain at Duke, although I had had numerous offers for other well-paying positions.

Between my apartment and Duke was an unused parcel of land that interested me. It was as close to campus as it was possible to live. It had large trees and was so overgrown with honeysuckle and poison ivy that I couldn't get in to look at it. I asked a real estate broker to find out if it was for sale. After examining official records, he replied that the property, though not on the market, might be for sale. It consisted of two city lots and had belonged to a Rufus Pickett, a member of a well-known Durham family. After his death the heirs had decided to sell the land at auction, but some of them had thought that the highest bid, $6,500, was too low. When they later agreed to sell, the prospective buyer had lost interest.

The broker thought the heirs might sell if offered the old price, or even a little less. I reasoned that if only one of them wanted more, the result would be lengthy negotiations; even worse, they might put the property on the market again, and I might lose it. I therefore offered $7,000 but told the broker, "Don't let anybody know that the buyer is a Duke professor, or the price will go up." My offer was accepted, and I became owner of a property I had not yet set foot on.

Apartment living had worn me down more than I realized, and planning for my new property gave me a mental boost. I contacted William Sprinkle, the architect who had designed our previous house. I especially liked his ability to design houses as his clients wanted them, rather than as he wanted to build them. I had no thought of ever marrying again and told him what I needed: a large living room with an adjacent study, a bedroom in one corner of the house and a guest room in another, bathrooms, kitchen, basement, and garage.

Sprinkle made some preliminary sketches and had the lot surveyed for placement of the house, driveway, entry to the garage, and steps to the main door. I thought about room arrangements and how livable and comfortable the house would be and left it to him to decide about details such as width of doors and height of kitchen counters. To make certain that the space was well utilized, I went over his plans with a black pencil and shaded all areas I considered wasted. Moving or reducing the blackened areas was like solving a puzzle.

In the early summer of 1964, after approving final plans for the house, I left for Arizona to study jackrabbits, another desert animal that had puzzled me for years. Unlike real rabbits, which live in burrows where they can escape from the daytime heat, jackrabbits are hares and don't have burrows; they remain above ground and are exposed to a terrific heat load during a summer day.

Camels, we knew, manage well if they have water, but there is no free water where the jackrabbits live. Moreover, because they are fairly small animals, they presumably have high metabolic rates and heat production. In addition, they have a greater surface-to-body-volume ratio than a large animal, and on a hot day their heat gain would be relatively much higher than for an animal the size of a camel.

An obvious feature of jackrabbits is their inordinately large ears, but what good are large ears in a hot desert? As long as air temperature remains below the animal's body temperature, the ears act as cooling devices that help the animal lose heat. But if the air temperature exceeds body temperature, we would expect the large ears to absorb heat from the hot air. How, then, can jackrabbits live where common sense says they cannot?

Two of my postdoctoral associates, Terry Dawson and Don Jackson, spent the summer with me, working on these problems at the Santa Rita Research Station, where Bodil and I had done our first work with kangaroo rats. Another colleague, Ted Hammel from Yale, joined us, along with Dave Hinds, a student at the University of Arizona.

In the extensive desert plains below the Santa Rita Station, kangaroo rats and jackrabbits were plentiful among the cactus and scattered desert shrubs. Above the station was a deep canyon with lush bushes and oak trees along its steep sides; in the wet season the dry creek at its bottom was a raging torrent that drained water from the Santa Rita Mountains.

We worked day and night. At night we hunted jackrabbits, relying on a technique perfected by Dave Hinds. Driving a pickup truck off the road, he used a spotlight mounted on the roof of the cab to scan the surroundings. When the strong beam of light hit a jackrabbit, it froze in place, and the rest of us tried to sneak up on it and catch it with our hands. Sometimes the strategy worked, but too often the animal sensed our footsteps and scooted off into the darkness.

Then Dave asked someone else to drive while he stood in the bed of the truck with a light rifle. When a jackrabbit was caught in the light beam, he shot over our heads, aiming beyond the animal. The sound confused it, and while it hesitated we could often grab it. The system worked fine, except for scraped knees and cactus spines we couldn't avoid in the dark.

We kept the animals in a large enclosure. During the day we measured their heat exchange under true desert conditions, meticulously recording air temperature, solar radiation, heat radiation from the ground, and the animals' body temperature. We later supplemented our field measurements with determinations of metabolic rates and water loss under accurately controlled conditions in the laboratory at Duke.

We found that on a hot day the jackrabbit reduces the heat load by exploiting every bit of shade. For example, a jackrabbit may sit under a small mesquite bush, carefully moving around the trunk to remain in shade as the sun moves over the sky, somewhat like a living sundial, thus reducing the heat load to a minimum.

We determined that as long as the air temperature remains below the jackrabbit's body temperature, the ears function as expected: metabolic heat is lost by conduction to the air, which obviates the need to use water for cooling by sweating or panting. Heat loss from the ears is aided by a high rate of blood flow; when the air temperature is below the jackrabbit's body temperature, the ears are pink and visibly engorged with blood.

However, when the air temperature exceeds body temperature, the ears become a liability. If the air is warmer than the blood streaming to the ears, the blood will return to the body warmer than when it entered the ears, thus increasing the heat load. But we observed that as soon as the air temperature increased above a body temperature of about 40°C (104°F), the blood flow to the ears was reduced, the blood vessels constricted, and the ears turned pale. The reaction is related to, but the opposite of, the vasoconstriction that in winter occurs in our hands to reduce heat loss. In the jackrabbit, of course, vasoconstriction serves to reduce heat gain.

We worked hard, but had occasional diversions. One Sunday we drove south to Nogales and crossed into Mexico, where I saw my first and last bullfight. It was disgusting to watch a man tease a furious bull in order to amuse the public. I knew that an experienced matador would probably escape unhurt, for a number of men on horseback were in the arena to divert the bull's attention. The terrified horses evidently were expendable. The bull gored one of them, apparently severing a major artery, for the horse collapsed within seconds and was quickly dragged from the arena. The jubilant public yelled for more. I had seen enough and left the gruesome display.

At the end of the summer I returned to a very active research group at Duke, bringing with me some ten jackrabbits in order to measure their metabolic rates and water loss under laboratory conditions. My research students were among the best I ever had, and with three postdoctoral associates I was surrounded by stimulating young scientists. With my house plans complete, I signed contracts with a builder, and construction began.

Confident that all was well, I decided to spend the month before classes started on a safari in Kenya and Tanzania (then known as Tanganyika) that Malcolm Gordon had asked me to join. Malcolm was knowledgeable, intelligent, and always fun to be with, and from our work together on crab-eating frogs, I knew he had a keen eye for animals and would be an excellent safari companion.

I flew to Nairobi, where I spent a couple of days with Dick Taylor. Dick had completed his Ph.D. at Harvard the previous year, and before leaving for Kenya he had visited me at Duke to discuss his far-reaching projects for research on African wildlife. Now he was in Nairobi, looking exactly as I remembered him. Tall and heavily built, he might seem clumsy as he moved about, but anyone who thought him big and bumbling would be mistaken; he was strong and agile and had a fine record in college athletics. Years later, his many friends and colleagues were saddened when he died from a heart attack at the age of only fifty-six.

We drove in Dick's jeep to Muguga, about an hour west of Nairobi, to the East African Veterinary Research Organization. Here Dick had established himself as a one-man research team to study the physiology of African mammals. As always it was stimulating and exciting to talk with him about his work. He was full of ideas and talked easily, joking and laughing; sometimes he might seem absentminded, but he was never inattentive. With animals as with people, he was gentle and kind, and he was very much in his element in the laboratory he had set up at Muguga.

Dick had completed a substantial volume of excellent research in a short time under primitive conditions. Before he went to Kenya, I doubted the wisdom of sending an inexperienced Ph.D. alone to Africa for a major research project; at Muguga I realized that he was an exceptionally gifted and productive investigator. What he had learned about African mammals—among them gazelles and wildebeest—and how they lived in the heat with very little or no water exceeded all previously existing information combined.

Returning to Nairobi, I was introduced to our guide, Bob Lowis, a lean Englishman who had been born and raised in Kenya. Our party included Malcolm and his wife and two other couples. We left Nairobi at dawn, driving south in two Land Rovers toward the Amboseli Game Park, our first overnight stop.

We soon left the paved road and followed tracks through the countryside. The African savannah, so familiar from pictures, became real for us. Acacia trees dotted the landscape, their thick branches spreading horizontally, their undersides clipped smooth a couple of meters from the ground, the level browsing animals can reach. I couldn't help wondering how an acacia seedling ever reaches a sufficient height to be out of the reach of browsers. On one long stretch we stirred up enormous clouds of red dust, which explained why the stripes of zebras in a nearby herd were black and pink! Time after time Bob pointed out animals we hadn't noticed—antelope, gazelles, giraffes, ostriches.

In the game park at Amboseli we got still closer to the animals. At one point we were even charged by a full-grown rhinoceros, which came toward us like a locomotive. After calmly moving the Land Rover a short distance away, Bob explained that rhinos have poor eyesight; when they come near enough to see how big a vehicle is, they invariably stop—except once, he added, telling us about one that couldn't stop in time. Its horn went right through the door and into the driver's buttock. The driver wasn't badly hurt, but the rhino had to struggle to get dislodged.

Bob also told us about some people touring in a Volkswagen who foolishly amused themselves by feeding oranges to elephants, oblivious of the consequences. The next time a Volkswagen came along, the elephants wanted oranges, and when they didn't get any, they turned the vehicle over to see if there were any underneath.

We came across lions almost daily, and it was always exciting to observe them. Beautiful giraffes browsed on the acacias, unhindered by the two-inch-long thorns. Even when galloping over the plains, their movements were slow and graceful. The zebras looked smooth and round and well fed, although in fact, like all other African game animals, they are exceedingly lean, with virtually no fat in their bodies.

We passed shallow salt lakes where thousands of flamingos stood in the water, brilliant pink as the sun shone on their feathers. Through our binoculars we watched them stretch their long necks down and, head upside-down, strain from the water small planktonic crustaceans containing the carotenoid pigments that account for the beautiful coloration of the birds' feathers. As we drove closer, the flamingos took to the air like a pink cloud lifting slowly from the water; Bob told us that between 100,000 and 200,000 flamingos resided at this lake.

On the plains of the Serengeti Game Park, we saw thousands of wildebeest migrating, moving in endless rows at a gentle gallop, one of the most remarkable animal migrations known to humankind. On one day in Serengeti we spotted twenty-eight species of mammals—lions, leopards, cheetahs, hyenas, jackals, antelope, gazelles, baboons and other monkeys, buffaloes, waterbuck, warthogs, and kitfox—almost every kind of large mammal that lives in East Africa.

Our three weeks in this remarkable country included

a visit to Olduvai Gorge, where the Leakeys had made their most important finds of human fossils. At the end of the safari I again stayed with Dick Taylor in his cabin, which felt cold and drafty. That evening I understood why I felt so wretched; I had a high fever. I was better the next morning, but for several days I felt miserably cold. I never found out what was wrong, but Dick and I agreed that I wasn't sick enough to suspect malaria. A few days later I gave a lecture at the Kenya Academy of Sciences; I again thought the building was unreasonably cold, although no one else complained.

From Nairobi I flew to Egypt, where I contacted the scientists who had shipped Egyptian sand rats to me at Duke. I wanted to arrange for a continued supply, since they were important for the study of kidney function and the causes of diabetes. I had become familiar with sand rats in Béni Abbès, where the French zoologist Francis Petter brought them to our attention. They live where the soil is highly saline, and they feed exclusively on fleshy saltwort and other juicy plants notorious for having sap with a salt content twice as high as that of sea water. No animal should be able to eat such salty plants, but it turns out that sand rats have kidneys extraordinarily good at excreting enormous amounts of salt.

At Duke we had trouble keeping sand rats alive. Despite growing fat and looking healthy, many died unexpectedly. A colleague in the Pathology Department, Don Hackel, found that the dead rats' pancreases had lost all their insulin-producing cells; evidently the animals had died from diabetes. A check of live animals showed enormously high blood-sugar concentrations and confirmed they were suffering from severe diabetes.

What was causing the diabetes? It seemed that the more we worked to improve the rats' diet, the more animals died. In nature these animals eat the fleshy parts of plants high in salt but low in calories, so we changed our approach and gave them a minute amount of ordinary rat food and a free selection of beets and spinach, plants of the goosefoot family, closely related to the rats' natural food though not as rich in salt. Kept on a limited ration of calories, the rats stayed well. With free access to food, they developed diabetes, a pattern strikingly similar to the development of the disease in diabetes-prone humans. Obese humans have a much greater tendency than lean persons to develop diabetes, and many patients with adult-onset diabetes can keep the disease under control with weight loss and a limited caloric intake.

We continued these studies for several years, until it became too difficult to import a sufficient number of animals. We also tried breeding the rats, but instead of mating, males and females fought and killed each other. I now realized that one important prerequisite for successfully domesticating a wild animal is its willingness to breed in captivity.

In October I attended a conference in East Germany, and on the way I stopped in Sweden to see family and in Denmark to visit my sculptor friend Niels. Since my divorce I had been uncertain about visiting him because his wife, Ellen, was Bodil's sister. The year before, when passing through Denmark, I had called him; he had sounded overjoyed to hear my voice and had immediately asked to see me. This time I knew I was welcome; Niels and I sat talking until four in the morning, when we heard the milkman rattle the bottles on his rounds.

Although the East German conference itself was interesting, spending a few days in a communist country was a memorable experience. Most noticeable was the pervasive atmosphere of sadness and depression; people in the streets appeared subdued and lifeless, gray and defeated. Nobody smiled; no person, young or old, looked directly at me, only past me, as if I didn't exist.

After the meeting, Karl Ullrich, a distinguished kidney physiologist who had worked at Duke some years before, invited me to drive with him back to Berlin. Before we were allowed to leave East Germany and enter West Berlin, our passports and the car in which we were traveling were carefully examined at a series of checkpoints. Once we had to drive over a giant mirror for the police to check that no contraband was hidden underneath. Then we had to leave the car so the seats could be lifted to ascertain that we weren't smuggling anybody out of the communist paradise it was forbidden to leave.

I returned home to find that construction had finally begun on my new house. Much of the heavy undergrowth had been cleared, so I could finally see my land, excavation for the basement had started, and soon the foundation was poured.

My laboratory bustled with activity. I had a handful of enthusiastic research students and three postdoctoral associates. Dave Hinds came from Arizona to measure the jackrabbits' responses to heat, supplementing the data from the summer's fieldwork. Peter Bentley came from the University of Bristol, England, and provided further stimulation. He started a study of the water loss from the skin of lizards and snakes. Until then it had been generally assumed that the dry and scaly skin of these reptiles was impermeable to water; Peter showed that the cutaneous water loss was much greater than had been thought.

Going to and from work I liked to pass by my growing house. I often met with Mr. Sprinkle, a remarkably meticulous person, who came to the building site almost daily to see for himself that every detail was correct. The builder was competent, the construction progressed on schedule, and on June 4, 1965, I was in my own house, no longer living in the apartment, where I had felt less and less at home.

I was preparing a lecture for the International Congress of Physiology in Tokyo in September 1965. I wanted to report on my student Gene Crawford's interesting studies on a tame ostrich we kept in the laboratory.

The ostrich is a typical desert bird that has occupied the Sahara Desert in historical times. Near Béni Abbès I had found numerous fragments of their eggshells, but ostriches are now extinct in the Sahara, exterminated by humans. Fortunately, they are common elsewhere, including the Kalahari and Namib deserts in southern Africa.

Although we had good information about desert mammals—the small kangaroo rat, the medium-sized jackrabbit, and the much larger camel—knowledge of desert birds was limited. Small birds can reduce heat stress by finding microclimatic pockets where they are less exposed, but ostriches are too big to find enough shade in which to shelter. We wanted to know how ostriches could tolerate the desert heat and manage their water balance.

A breeder had shipped us an ostrich chick that weighed 14 kilograms (29 pounds). It grew up and became the subject for Gene's doctoral dissertation, eventually weighing 100 kilograms. When it neared adult size, it grew white wing feathers in distinct contrast to its overall dull grayish-brown color. We therefore assumed it was a male that gradually would change its color to the characteristic black of male ostriches. We called the bird Pete in honor of my good friend Pete

Scholander, the widely recognized animal physiologist who was my scientific hero.

As an adult, Pete stood much taller than I and weighed half again as much, but he was very tame and fairly easy to handle. One morning Gene walked into my office, noticeably upset.

"There is something funny with Pete's behavior," he announced. "He came against me with outstretched wings as if to knock me down. And I don't want to be raped by an ostrich."

We had never thought of that kind of trouble, but it was springtime, and thus it might be reasonable for a mature male to become troublesome. The huge bird might be impossible to control if in a mood to mate. He had been handled by humans all his life and was probably imprinted on us; we knew that the zoologist Konrad Lorenz had raised geese that were imprinted on him and that later had erroneous ideas about humans as suitable sex objects.

I immediately went with Gene to Pete's pen. The moment Pete saw me, he rushed toward me with head lifted high and wings outstretched, looking very excited. I patted him gently on the back and said, "Now, now, Pete." Instantly his legs folded under him, he stretched his long neck along the ground, and in a soft ostrich voice he muttered endearing words.

In a flash I realized that Pete was a female, which we should have known because "he" had never grown the male coat of black feathers. "I never saw a female who fell for me that fast!" I laughed. We continued to call her Pete even though she started producing large, beautiful eggs. They were unfertilized, of course, so we couldn't incubate them, but I became an expert at preparing ostrich-egg omelets.

I gave my research students a problem. An ostrich egg weighs one kilogram, and one kilogram of hen's eggs is sixteen eggs (this I had learned from Mother). Then, if it takes ten minutes to hard boil a hen's egg (also learned from Mother), how long will it take to hard boil an ostrich egg? None of the students offered a solution, which involved relative surface areas and volumes as well as the distance from the surface to the center of the egg. I myself did not know precisely how to do the calculations, but I made an estimate and boiled one of Pete's eggs for forty-three minutes; it was nearly perfect. The yolk was firm except for a small semisoft spot at the center. A few minutes more would have been better.

Gene found that the ostrich is remarkably tolerant of high temperatures. It fluffs up its feathers in the heat, as other birds do in the cold, thus increasing its surface insulation. When Pete fluffed her feathers, the thickness of her coat increased from a couple of centimeters to about ten. The increased insulation greatly reduces the heat transfer from the surrounding hot air, and is even more important than the fur is for the camel.

Even in temperatures as high as 51°C (124°F), Pete could maintain her normal temperature of 39.3°C (102.7°F) for hours. At this temperature the feathers were insufficient to maintain a normal body temperature. Under such conditions, the ostrich needed to increase evaporation by panting, much as dogs do in hot weather. At high temperatures Pete increased her respiratory rate from 5 breaths to 45 per minute. This does not seem very high by comparison with the rate for a dog, which may pant at 300 or 400 breaths per minute. Nevertheless, at this rate the ostrich kept her temperature normal, even after hours in high temperatures. As far as we could judge, Pete was extremely well adapted to desert life; we had exposed her to temperatures several degrees higher than found in even the hottest deserts.

Before I left for Tokyo, Terry Dawson suggested that when I returned and before he left we ought to have a big party to celebrate my fiftieth birthday. I agreed and suggested that if Pete laid any eggs in the meantime, we could serve ostrich omelets. Otherwise we made no firm plans. During my absence Terry and the others decided that the party should be a culinary event based entirely on gastronomic products of North Carolina. When I returned everything was ready for a feast on September 25, 1965, the day after my birthday.

All thirty-seven people who came helped make the party as true to local color as anybody could wish. Astrid and her husband, John McHugh, came from Washington to help. Bent, who was then in Durham, caught frogs at the ponds on Duke's golf course and contributed thirty frogs' legs. My faculty colleague Joe Bailey had caught and prepared a large snapping turtle. Our laboratory technician, Judy Scarff, returned from a coastal visit to her parents with shad roe and a huge amount of fresh shrimp; the shrimp gumbo fully deserved the universal praise it received.

People kept coming with more surprises. One of Joe Bailey's students, Dave Osgood, was a devoted hunter; he and his wife brought dove's breasts, each about the size of a soup spoon. They were my favorites. Three dozen should have been enough for everybody to taste, but there was so much else to eat that I didn't feel guilty about consuming more than my share. Dave also cooked a real Brunswick stew, using squirrel and wild rabbit, as the pioneers had, and home-grown corn, lima beans, and tomatoes.

I had roasted two locally raised geese, but only one was eaten—there was too much other food, including omelets I made from Pete's eggs, locally produced if not a typically North Carolina staple. I also cooked an opossum stew, but the finished product was greasy and gristly; despite hours of cooking, the little meat on the bones was as tough as truck tires and tasted dreadful.

Gene Crawford had promised to provide white lightning, a powerful home-distilled corn whiskey that is as clear as spring water. But Gene's source had dried up. I felt that white lightning was needed, so I found a reliable supplier. When obtaining white lightning, one must avoid the stuff that is contaminated with lead, produced when people use car radiators instead of clean copper tubing for condensers; car radiators are lead soldered and produce a rather toxic brew. My source provided high-quality stuff, brilliantly clear and colorless, smooth and guaranteed to be aged less than three weeks.

It was indeed a memorable party. All my friends and colleagues were there, everybody had contributed to its success, and everybody had a good time.

The Unusual Animal

GENE CRAWFORD'S stud-
ies of Pete the ostrich were interesting, but laboratory experiments with
one bird raised in captivity didn't tell us much about how ostriches
function under natural conditions. The place to go, I decided, was South
Africa, where ostriches are raised commercially on large farms. I
arranged for permission to work at a ranch owned by the Onderstepoort Veterinary Research Institute, about ten miles from Pretoria in
the Transvaal. Ostriches could be shipped to the ranch in time for our
arrival.

I was sorry that Gene Crawford was unable to join my research
group; he had just been appointed to a new faculty position at the
University of Kentucky and couldn't leave. Instead the team consisted
of John Kanwisher from Woods Hole Oceanographic Institute, a physicist turned physiologist; Bob Lasiewski from the University of Cali-

fornia at Los Angeles, an expert on the physiology of hummingbirds; Jerry Cohen from the University of Kentucky, an associate professor of medicine who knew more about bird respiration than most physiologists; and Bill Bretz, one of my graduate students at Duke.

The ranch we worked at, known as Kaalplaas, was ideal for our project. It was some two miles from Onderstepoort, the internationally known research institute and veterinary school where tropical livestock diseases were studied and new vaccines developed. It had about one thousand employees and produced hundreds of thousands of vaccine ampoules each week, with a yearly production amounting to some eighty million.

The researchers kept a formidable number of animals for study and the production of vaccines. I copied their list for just one day; on September 22, 1966, they had 301 horses, 55 mules, 930 cattle, 2,666 sheep, 201 goats, 61 pigs, 162 dogs, 11 jackals, 6,698 guinea pigs, 1,012 rabbits, 56,314 mice (an unusually low number, they said), 44 cats, 295 hamsters, and 570 breeding rats. Our 20 ostriches and other "exotic" animals were not listed as part of the regular stock.

Kaalplaas was less hectic and more like an ordinary farm, with a few outbuildings and many horses and cattle. The nights were beautifully clear and pleasant, at times so cool I needed a sweater. When the moon waned and the Southern Cross shone clear and brilliant, I taught John Kanwisher to locate true South by the stars, in the same way that almost everyone in the Northern Hemisphere knows how to find true North by Ursa Major. South is in a conspicuously dark and empty part of the sky, with no marker as obvious as the North Star. Instead you must start from the Southern Cross, and with the aid of two stars known as the "pointers" and a very complex set of rules about diagonals, straight lines, and intersects, you can arrive at true South. The complex method worked, and John soon taught the others.

The ostriches at Kaalplaas were big, strong, and fierce. The foreman, Mr. DeLange, warned us that an ostrich can eviscerate a man with a single stroke of its foot. Big ostriches weigh more than 100 kilograms— more than a hefty football player. Most of the bird's muscle mass is concentrated in the legs, and the foot ends in a sharp claw. Imagine a huge man hitting you with a sharp crowbar bearing a spike at its end, using all his strength, and you can appreciate the kick of an ostrich.

Our ostriches were kept in a large fenced paddock. We used horses to round them up and steer them toward the chutes, where we singled out individuals we wanted to study. Ostriches, unlike horses, kick only forward. When we moved an ostrich from the chute, one man held onto each wing while a third grabbed hold of the tail feathers; this way we could handle even the biggest ostrich safely. The maintenance department at Onderstepoort had constructed a temperature-controlled room for us and had built restraining devices for the birds so we could study their reactions to high temperature.

I loved being able to spend all my time with the animals and in the laboratory, free from the bureaucratic office work and administrative duties that took so much time at home. I could do what I liked most—work with interesting animals and a marvelous group of collaborators.

We were amazed that we could keep the ostriches for hours at 55°C (130°F), while they panted at about fifty breaths per minute, ten times their normal rate. During these experiments one of us had to be in the hot room with the bird to make certain that there were no accidents. If the air is dry, such a high temperature is not a serious problem for a human who has enough water to drink.

We expected the birds to tolerate high temperatures, but we were surprised that their panting didn't affect the alkalinity of their blood. In earlier studies one of my students, Bill Calder, had found that all birds he tested inevitably lost excessive amounts of carbon dioxide when they panted. As a result of loss of carbon dioxide, the pH of their blood increased, becoming so alkalotic that it would be lethal to humans.

The respiratory system in birds is far more complex than that in mammals. Not only are the lungs structured differently, but also birds have a number of large, thin-walled air sacs that have no counterpart in mammals. Like other terrestrial vertebrates, birds breathe in and out, inhaling through the mouth or nares and exhaling before taking the next breath. But how air flows between their lungs and air sacs was poorly understood. Did it enter the lungs first and then flow to the air sacs, or vice versa?

To find out more, we let an ostrich take one breath of pure oxygen and, using an oxygen electrode, followed the movements of the oxygen in the respiratory system. We found that the oxygen concentration increased sharply in the large abdominal air sacs and only later elsewhere in the respiratory system. We could therefore conclude that on inhalation the oxygen flowed directly to the large air sacs in the abdomen, and only later moved to the lungs.

After we returned home, Bill Bretz designed a minute air-flow probe that he placed in strategic places in the passageways to the lungs and air sacs of domestic ducks. He confirmed that inhaled air first flows directly to the large abdominal air sacs and then flows through the lungs and into air sacs located anterior to the lungs. The air in the anterior sacs, having lost much of its oxygen content in the lungs, is eliminated on the next exhalation. Thus our successful studies of the heat tolerance of ostriches eventually resulted also in the solution of a major unsolved problem in bird respiration.

AFTER SOME weeks of working all day every day, I asked Mr. DeLange if he would let me ride a horse that had just been used to round up an ostrich. He hesitated. "It is a little too fast for the professor; I will give the professor another horse," he said. I wasn't interested in a sedate horse and declined, saying that I really was too busy.

One Saturday afternoon when no workmen were around and I needed to pick up a key from Mr. DeLange, I quickly saddled a horse and rode up to his house. DeLange saw me arrive on one of the liveliest horses and evidently changed his opinion of what was suitable for "the professor." The next time I asked, he gave me his own horse, Jock, with his own saddle.

Jock was probably the finest horse of the 100 or 150 at Kaalplaas. He was strong and eager to run, with incredible stamina, and was the fastest horse I had ever been on. After cantering for miles across the plains, he was still eager to go as fast as I would let him. On a halfday inspection trip with Mr. DeLange we covered mile after mile of the vast areas that provided feed and pasture for the thousands of animals at Onderstepoort.

Under the existing system of apartheid, all the laborers on the farm, as everywhere else in South Africa, were black. All whites, including my group of scientists, were called "Europeans." The blacks were called

"Africans," or simply "black bastards," or some other derogatory term. In between were "coloureds," persons of mixed native and European origin. All races were segregated by law.

The natives had to carry identification cards and work cards signed each month by their employer to show legitimate employment; the whites did not need such cards. Post offices, banks, and other public and semipublic businesses had separate entrances for whites (*blankes* in Afrikaans) and nonwhites (*nie blankes*), and there were separate buses and trains for Africans. The quality of the segregated schools differed drastically. Stores, on the other hand, were not segregated; it was perfectly legitimate to accept money from nonwhites.

Even then there was strong opposition to the policies of the Afrikaans-speaking whites. For example, the faculty at the University of Witwatersrand in Johannesburg successfully resisted a government order to dismiss all black students, even when the government threatened withdrawal of financial support. However, in that highly industrialized country with an excellent scientific establishment, a strong military, and incredible natural resources, it was not then clear whether nationwide change could be brought about through moral pressure from the rest of the world.

On MY WAY home from South Africa, I stopped in Denmark, where Niels met me at the airport. To my immense delight I also saw Mimi, who was in Denmark for a year, working with the electronics firm Bang & Olufsen.

I returned home a couple of weeks before Christmas, having been away from Duke for the entire fall term. These long absences, which permitted me to carry out research projects in other parts of the world, were possible because I had received a lifelong appointment as a career investigator of the National Institutes of Health (NIH). I could spend more time on research while remaining in my university position because part of my salary was paid by the NIH. This reduced my teaching obligations to a course in animal physiology each spring term, leaving the fall free for research.

ONE MORNING soon after returning home I received a call from David Rall of the National Cancer Institute. I knew him well from summers at the Mount Desert Island Biological Laboratory in Maine, where he had studied an interesting phenomenon known as "the blood-brain barrier." This barrier exists because the capillaries that supply blood to the brain are less permeable than other capillaries, thus forming a "wall" that blocks certain substances from moving from blood to brain.

"Knut, I need a title for your lecture," Dave said. At first I couldn't think of what he was talking about. Then I remembered. A year before, in September 1965, he had talked me into giving a lecture at a meeting on comparative pharmacology in Washington in January 1967. It had been difficult to turn him down, for I felt obligated to the National Institutes of Health for my research funding, and also for their policy of supporting research on fundamental problems in physiology.

Now, a year later, Rall was asking for the title of a lecture I had forgotten all about. I could only say, "But I can't give you a title on the spur of the moment. I don't even know what you want me to discuss."

Dave was insistent. "The program has to be printed, and we must have your title," he said. "It doesn't matter that you haven't prepared your lecture yet." I told him that I knew nothing about pharmacology and asked what he wanted me to talk about.

"You can talk about anything you want. Camels, or some other unusual animal you have worked with," he added casually.

I had an idea. "What about calling my lecture 'The Unusual Animal'?" I asked. That would give me time to contemplate what to talk about. Dave agreed at once, then shocked me by saying that my lecture would be an after-dinner speech at the evening banquet.

I had given an after-dinner speech only once before, years earlier at a meeting of the Sigma Xi Society. Mimi, in high school at the time, had gone with me, saying she would enjoy an evening at a scientific meeting. At the banquet I talked about water and camels and sea gulls, and on the spur of the moment I continued discussing other aspects of thirst, about goats that drank as much as one third of their body weight if a certain part of their brain was stimulated with a weak electric current, and about how rats, if they were made neurotic by contradictory training signals, would turn to drink and rapidly become alcoholics. The audience had liked my talk, and Mimi had been delighted that science was so much fun.

I remembered this talk after Dave Rall's call. Perhaps I could talk about overeating, rather than drinking. I would describe sand rats and how they develop diabetes when they eat themselves fat, as some humans do. The sand rats had unexpectedly become a useful model in research on diabetes, and I could also discuss other animals relevant to the solution of specific physiological problems.

The meeting ended with an elegant banquet, and when the time came for me to appear at the podium to give my lecture, people were still sitting at the tables, full and satisfied and sleepy after ample cocktails and good food. Nobody wanted a scientific lecture with boring slides of graphs and tables. But I got off to a good start: my first slide was a cartoon of an immensely fat person with a caption underneath, "Obesity—It's Incurable, So Relax and Enjoy It." The slide brought the house down in laughter.

Then I described how sand rats develop diabetes when they eat too much. I showed only one slide with data, a growth curve for diabetic sand rats. My other slides showed engravings of exotic animals, copied from centuries-old natural history books. I used these slides when I discussed other animals that in some way or another have been important in physiological research.

One example I mentioned was the guinea pig. Early in this century two Norwegian physicians, A. Holst and T. Frölich, discovered that guinea pigs are the only animals besides humans that get scurvy. The discovery provided a research tool for finding the causes of this ancient scourge of sailors. Holst and Frölich reported that guinea pigs, unless fed on cabbage or other fresh vegetables, developed symptoms remarkably similar to human scurvy.

At that time nobody imagined that human disease could be caused by nutritional deficiencies. After Pasteur's discoveries, the search for causes of human illness was concentrated on searches for microbes, and it was known that scourges such as cholera and tuberculosis were caused by bacteria. Scurvy, like tuberculosis, was associated with poor living conditions and afflicted sailors at sea who lived under miserable hygienic conditions. Why shouldn't scurvy also be caused by bacterial infection?

Research has since revealed that scurvy is caused by lack of vitamin C, or ascorbic acid. Other animals, such as rats, cats, and dogs, synthesize their own vitamin C and don't need it in their diet. The crucial discovery that disease could be caused by a nutritional deficiency was based on the perceptive observation of an animal that, like humans, is uniquely susceptible to lack of ascorbic acid.

I then talked about the role that two other animals—the common frog and the angler fish—played in clarifying kidney function. In the 1920s and 1930s, two major theories prevailed about how urine is formed in the kidney. Either waste products were filtered from the blood as it entered the kidney, or wastes were secreted in the renal tubules in a manner similar to the secretion of saliva by the salivary glands.

In the 1920s the pharmacologist A. N. Richards proved that filtration occurs in the frog kidney, as had long been surmised. Then, in 1928, came the remarkable discovery that filtration is not necessary for urine formation. A few animals exist whose kidneys have no filtering mechanism; their urine must therefore be formed exclusively by tubular secretion.

The most important animal for these studies was the marine fish *Lophius piscatorius,* the angler fish. One never sees the whole fish in markets because of its ugly appearance—behind a broad, flat head with a wide mouth stretches a small body. In France this *lote de mer* is considered a great delicacy; American fish markets sell the fillets under the name "monkfish."

The angler fish is of special interest because its kidney lacks a filtration mechanism. The frog is important because it helped demonstrate the importance of the filtration process. Thus two seemingly insignificant animals helped resolve a long-standing controversy about kidney function, and we can say that both opposing groups of scientists were right.

In my speech I also discussed the squid, which has proved essential to our understanding of nerve function. Today almost all we know about how a nerve impulse is conducted along a nerve fiber derives from the study of squid. This animal, no longer than a human hand, has nerve fibers, or axons, so inordinately large compared with those of other animals that they are called "giant axons." A single giant axon may have a diameter of almost one millimeter with a cross-section many thousandfold greater than a single mammalian nerve fiber. What accounts for the large size of the squid's axons? Squid swim by jet propulsion, which means that their muscular body wall must contract very quickly, forcing a jet of water out through a funnel-shaped opening, the siphon, propelling the animal in the opposite direction. Unless the whole mantle contracts at once, a strong jet cannot be produced. Because thicker nerves conduct impulses faster, the giant nerve fibers conduct so fast that even the most distant parts of the mantle contract without appreciable delay.

The size of the giant axons is helpful for research. Neuroscientists can insert a fine electrode into an axon and place another electrode on the surface and study electric changes. A light electric shock at one end of the axon causes a nerve impulse to flow along the axon, and when the impulse passes the electrodes, the change in electric potential across the nerve membrane can be recorded.

It would be a mistake to believe that the nerve impulse is like electricity flowing in a piece of copper wire. Inside a nerve fiber is a much lower concentration of sodium than outside, and the English physiologists Alan Hodgkin and Andrew Huxley showed that at the instant a nerve impulse occurs, the permeability of the nerve membrane changes. This permits an instantaneous flow of sodium ions across the membrane from the high concentration outside to the low concentration inside. Because each sodium ion carries a positive electric charge, this flow of ions is picked up by the electrodes as a change in the membrane potential, a change that was not understood until the squid's giant axons permitted direct observation. Hodgkin and Huxley's discoveries, now part of classical physiology, were honored with a Nobel Prize in 1963.

How does the permeability of the nerve membrane suddenly change? It turns out that sodium channels, or gates, exist in the membrane, and when a nerve impulse passes, these channels open for an instant, permitting sodium ions to flow through the membrane and cause an electrical change. How these gates work was clarified with the aid of a highly toxic substance, tetrodotoxin, which attaches to and blocks the sodium channels.

Because tetrodotoxin is such a potent poison, a hundred times as toxic as curare or strychnine, I thought it might be of special interest to the many pharmacologists in my audience. Tetrodotoxin is found in the organs, especially the liver, of puffer fish, so called because they blow themselves up to nearly a sphere when pulled out of water.

In the Orient the lethal toxicity of the puffer fish has been known for thousands of years. In the Western literature one of the first reports was by Captain James Cook, when his ship was anchored at New Caledonia in 1774. A new fish brought on board was given to the naturalist to be sketched and described. He took so long that the fish could not be prepared for a meal, and only a small piece of the liver was eaten. Cook recorded: "About three to four o'clock in the morning we were seized with most extraordinary weakness in all our limbs attended with numbness of sensation. . . .We each took a vomit and after that a sweat gave great relief. In the morning one of the pigs which had eaten the entrails was found dead."

One of the earliest articles about tetrodotoxin in a modern periodical was published, of all places, in *Playboy* magazine in 1964. In a story by Ian Fleming, James Bond visited Japan and was told about the blowfish, or *fugu,* which is used for murders and suicides because its poison brings instantaneous death. However, if correctly prepared, fugu is considered a delicacy and is eaten by the Sumu wrestlers because of its reputed strength-giving properties.

In Japan there are restaurants that specialize in fugu, recognizable by a model of the fish displayed outside. The fish is prepared by government-licensed cooks who go through a lengthy education. It has been said that, upon completing his education, a cook is licensed only after eating his own prepared dishes. I have eaten fugu, but I admit that I paid less attention to its taste than to any feeling of weakness in my limbs.

Japanese friends told me that if one eats fugu among friends and not in a restaurant, there is always one man who tries the fish first. If he falls over dead in a minute or two, the others go home after agreeing on the next time to meet. A true gourmet is supposed to eat just enough to feel the lips tingle.

In my lecture I showed an old engraving of a puffer fish, blown up to a sphere and covered with wounding spines. The pharmacologists in the audience, who knew the extreme toxicity of tetrodotoxin, were clearly relieved that fugu was not on the evening's menu. They listened with renewed attention to my descriptions of other animals that have served science, from fireflies to clawed frogs. I said little about my own work except for the sand rats, but preparing the talk had been great fun.

In JANUARY 1967 I went to San Diego to meet with Pete Scholander and the administrators of the Scripps Institution of Oceanography at La Jolla. The purpose was to discuss plans for an expedition to the Amazon River in Brazil in May and June that year, on which I was to be chief scientist.

Scholander was in charge of the scientific programs for a research ship that had been designed and built according to his specifications. The rationale was that since the world is full of unsolved scientific problems, instead of having individual scientists travel with often inadequate equip-ment to faraway places, it would be more effective to equip a ship with modern laboratory facilities and permit scientists to live and work on the ship while studying interesting animals.

Before this ship was built, there had been many oceanographic research ships that could collect and preserve animals, but not one had been equipped primarily for the study of living animals. For this undertaking Scholander had received funding from the National Science Foundation. When the ship was launched, he named it the *Alpha Helix,* after Linus Pauling's term for the structure of protein molecules.

To fulfill its mission, the *Alpha Helix* was assigned to a team of investigators who under the direction of a chief scientist would carry out interesting research projects. One of our main scientific objectives on the Amazon cruise was to study lungfish and other air-breathing fish. We were especially interested in the electric eel, known for its ability to produce electric discharges of several hundred volts, which it uses to stun or kill its prey. Curiously, its gills are inadequate for breathing, and the fish will drown if denied access to air.

Visiting Scripps to plan the expedition also gave me the opportunity to see Bent, who was working in Scholander's laboratory. Bent had started college at Duke, where he registered for an advanced chemistry course after tests showed that he had an excellent basic knowledge of chemistry. Chemistry was his best subject, and he made top grades. Then, in his typical manner, having shown he could readily handle the subject, he failed in chemistry. Not wanting to continue in college, Bent then secured a job as laboratory technician for Dan Tosteson, the professor of physiology at Duke. He worked in the lab for two years and became one of Tosteson's most skilled and dependable research workers.

In 1966 Scholander asked Bent to be a crew member on the first cruise of the *Alpha Helix,* and after the cruise he asked Bent to remain in his research laboratory at Scripps. Clearly, Scholander was impressed with Bent's competence; he described him as "a brilliant young student" in the *Annual Reviews of Physiology.*

When I visited Scripps in January 1967, Bent told me that he liked Scholander and would continue working with him, but he also wanted to finish college. He wanted no financial help from me; he would pay for his own education. I respected his wish for independence and was pleased with his plan. Bent completed his studies at the University of California at San Diego while continuing his work with Scholander, and after receiving his bachelor's degree he went on to graduate studies at the Massachusetts Institute of Technology.

Our discussions of the Amazon cruise went well. Captain Faughn, one of the most trusted skippers in the fleet of Scripps research vessels, would be in charge of the *Alpha Helix.* We settled the details about operating procedures; the ship would leave home port and sail to Manaus, where the scientific team would come aboard. The ship would remain on the Amazon for about half a year, with a number of scientific teams working consecutively for shorter periods. Mine would be the second team.

The *Alpha Helix* had cabin space for ten scientists, so I had invited nine others: the Norwegian physiologist Kjell Johansen and his colleague Claude Lenfant; Johan Steen from Oslo and his student Stephen Turitzin; Tom McManus and David Allen, both physiologists from Duke; Peter Hochachka, a biochemist from Vancouver whom I had known since he was a student at Duke; John Machin from Toronto; and Dennis Bellamy, a biochemist from Sheffield. We also had two Brazilian scientists on board, Luiz Junqueira and Jorge Petersen, from São Paulo. Later I learned that several other scientists, as well as two physicians from the University of California, would join us. The increased number would not cause undue hardship, however, because Scholander had planned for excess personnel to live in a shore camp while working on board the ship. The *Alpha Helix* would remain at anchor on the Río Negro, a tributary to the Amazon.

We stayed in Belem for a couple of days before flying to Manaus, more than 1,000 miles upriver, to join the *Alpha Helix*. The trip to Manaus gave us a fine view of the Amazon from the air. The world's greatest river was in flood, and we saw endless bodies of water, tributaries, channels, lakes, flooded land, almost as far as we could see. It was like an immense, leisurely flowing lake spreading endlessly over the land. On arrival at Manaus, we saw oceangoing ships from Liverpool and Bremen unloading and loading freight.

The *Alpha Helix* remained in port at Manaus for a few days before sailing upriver to its station on the Río Negro. At this time I established the routine of holding a research conference every morning at eight. At the conferences we went through the plans of each investigator, problems and methods, and coordination with other investigators. In this way we could ensure that scarce resources were utilized effectively. Lungfish, for example, which were in extremely short supply, could first be used for respiration studies by Johansen's group, then Steen could secure blood samples, and finally the biochemists could have samples of muscle and liver. Everybody came on time to these meetings; rarely was anyone as much as two minutes late. Captain Faughn always attended our conferences, and I appreciated his wish to be kept informed about our work. I asked each scientist to describe his plans, and we assigned working space in the laboratories. Everyone wanted more space than there was, but with some diplomacy all conflicts were resolved.

We had sent lists of the instruments and supplies we needed to Scripps, and were told that everything would be placed in the ship's hold in San Diego, waiting to be unpacked on our arrival. We found chaos. Our supplies were inadequately marked and scattered in several places, and our packages and crates had been opened by the preceding team, presumably because they were looking for items they couldn't find. A less generous soul suggested that they had hunted for equipment they had failed to order and subsequently discovered they needed. I was terribly disappointed that none of the equipment for my own research could be found. I had planned a study of the osmoregulation in the freshwater sting ray said to be common in the Amazon system and also the acid-base regulation of air-breathing fish.

I set up a series of evening seminars consisting of informal lectures presented by each investigator. We met every other evening at nine-thirty, which was the earliest feasible time. We all worked late, often after seminar time, and frequently even past midnight.

After food and other essential supplies were loaded, the ship started up the Rio Negro, continuing all day. We anchored for the night and got under way again early the next morning, reaching the location of the shore camp around noon. My first job was to see how the camp was arranged. Six people had to stay there, and I knew I had to be one of them, at least to begin with, to avoid the impression that the camp was an inferior place. Two others from my group, the two physicians, and Brian McNab, a zoologist from Florida, joined me at the camp.

Life on shore had some advantages and some disadvantages. The ship was anchored a few hundred meters out in the river, and to reach it we had to use a canoe or a boat. Most of the shore people worked on board but had lunch and dinner in the camp, so there was a great deal of moving back and forth. The camp had two small laboratory sheds, which provided extra working space. Each of two sleeping huts had room for six, and a somewhat larger hut housed the mess hall and the kitchen, where the cook, Baxter O'Brien, lived in a tiny cubbyhole.

Baxter was a huge, strong Australian with a tremendous black beard and hair in curls down his neck. His colorful language included more select profanity than I had ever heard. His linguistic talents undoubtedly appealed to Scholander, but the major reason he had been hired soon became clear. He cooked with skill, courage, and fantasy, keeping the inhabitants of his little kingdom supplied with a variety of unique food. Any unusual raw material inspired him, and he had an imaginative disregard for standard preparation methods. Whether it was an exotic fish or a turtle or a snake, eating at his table was never dull.

Baxter once wanted to bake an anaconda that was too big for his small oven, so he used the oven on board the ship. Some crew members objected to snakes being cooked on board and demanded that Baxter be barred from the mess and that the oven be sterilized before being used again. How sterilization would help was never discussed.

The Río Negro was in flood, so we could canoe for miles among the trees of the jungle, where philodendron climbed toward the dense canopy and a keen observer could always spot orchids. The clear, coffee-colored water was good for swimming and washing, and Baxter used it for cooking, for washing dishes, and for drinks. He never boiled the water, but no one suffered ill effects.

Baxter arranged things at the camp to his liking. He cut some pieces of tree trunks that had orchids growing on them and placed them outside the cabins. He kept a few animals as pets. Two young ocelot kittens roamed freely and playfully bit at our toes when we sat at the table.

A tame tapir also lived around the camp. At one time it broke out of its cage and disappeared; it later returned and gained the privilege of coming and going as it pleased. It slept away from camp but came in to be petted and fed like a dog, and it loved to swim with us.

At the market in Manaus, Junqueira had bought a large basket of Brazil nuts, and Baxter taught me how to crack them: put the nut against a wooden doorjamb, place your left thumb over the nut, and with your right fist hit the thumb as hard as you can. The trick is to hit hard enough to  break the nut; if you fail to break it, it hurts. But it is worth the risk; fresh Brazil nuts are infinitely better than the stale nuts available in U.S. grocery stores at Christmastime.

A capybara we named Cappy became a gentle and well-behaved house pet. Capybaras are the world's largest rodents, the size of a small pig, and semiaquatic. They live near or in water and have webbed feet, suitable for swimming. Cappy lived in the kitchen and around the dining table. He liked to eat crackers, bananas, sugar cane, mosquito nets, shirts, and other valuables.

The biochemists on board were interested in conducting experiments on Cappy, but those of us in the shore camp wouldn't dream of sacrificing our good friend. To solve that problem, I reported to those on the ship one morning that Cappy's cage had been found empty (which was quite true, as he hadn't been in it for days); he was nowhere to be seen at breakfast time (also true, since he was in the kitchen); and someone had seen him swimming in the river early in the morning (Junqueira had dragged Cappy along when he took his morning swim, but Cappy promptly freed himself and went directly for the shore). I couldn't help that the biochemists misinterpreted my truthful report. They didn't ask again for Cappy, who joined us at the dining table for every meal. Once he jumped up on the table; the effect was like a pig landing among the plates.

Despite the immense stretches of flooded jungle there were almost no mosquitoes. Malaria was not a problem, but we nevertheless used mosquito nets at night. Unfortunately, the nets couldn't keep out the obnoxious pium flies, tiny black flies so small they are hardly noticed. One cannot feel their bites, which develop into tiny purple blisters of blood. Scratching leaves a swollen, itching spot, much like a chigger bite. The pium flies were everywhere, especially on quiet days when we worked on the deck of the ship, and insect repellents didn't seem to faze them.

We had all heard about the piranha, a voracious small fish written about by many travelers. If a party has to cross a river, common knowledge has it that one first drives a horse into the water. A large number of piranhas immediately attack, each nibbling with its razor-sharp jaws, and in minutes the horse is reduced to a clean skeleton and the fish are satiated. Then it is safe to cross. We concluded that the stories were flagrant exaggeration, for we swam safely in the river, both at the shore and around the ship in midstream.

I stayed at the camp for the first couple of weeks, working on board all day but taking my meals and sleeping on shore. The huts were open and screened on all sides so that the breeze could flow through the room. The nights were so cool that I needed a sheet over me as I listened to the sounds of the jungle and an occasional tropical downpour that drummed on the tin roof.

One of the scientists in the land camp was Johansen, but when a bunk became available on the ship, he wanted to move back because he didn't get much sleep in the camp. At two-thirty every morning a flock of chickens with more than the usual number of roosters became active, crowing and running around among the poles that supported our huts.

There is no way to sleep with a cock crowing under your bed, but when Steen moved to the camp things changed abruptly. When dawn broke, he grabbed an oar and chased the biggest cock, swearing revenge. The rooster readily outran him, but Junqueira caught it by hand, and Baxter got another with his rifle. The next morning only one rooster crowed timidly. We encircled him and slowly closed in on him, and when he made a last dash for freedom, Junqueira caught him. We had baked rooster for supper.

One evening McNab caught a rare vampire bat, which he caged with a rooster overnight. The next morning the cock looked as healthy as ever, but there were droplets of blood around his anus, and the body weight of the bat had increased from thirty-six to forty-five grams. McNab reported that the vampire had drunk one-quarter of its body weight in blood.

McNab, a tall and lanky man, had installed his own equipment in one of the huts, where he studied the metabolism of a number of unusual tropical animals. Several times a day he went into the jungle to check his bat nets, which are known as mist nets because they are made of black thread, so thin they are barely visible even when you know where they are. With tremendous energy McNab would dive in among the trees and the strong climbing vines called lianas and always come back with something exciting. He caught all sorts of animals with his bare hands—wild rats, frogs, snakes, giant tarantulas—whatever he could see, and that was plenty. One day I saw him catch a bat on the wing with an ordinary, small butterfly net. He looked like a cartoon zoologist, tall and stooping, butterfly net in hand, nearsighted and wearing glasses, enthusiastic, and talking excitedly about everything he saw. He was more knowledgeable about field zoology than any of us, and unlike the other non-Brazilian scientists, he spoke reasonably fluent Portuguese. At an evening seminar McNab told us fascinating stories about his studies of rare and little-known tropical animals.

After a couple of weeks I moved back to the ship. Officially, the reason was that the chief scientist ought to be closer to the activities on the ship, but in truth, camp life had worn thin. Though informal and charming, it had less attractive aspects, not the least of which were the pium flies.

The cabin assigned to the chief scientist had only one bunk, and I was pleased with my solitary comfort. The day I moved in, I found a small glass containing an orchid in my cabin, and as long as I remained on the ship, there was a fresh orchid every day. However, I never found out who put them there.

Being alone in my cabin gave me a luxurious feeling of privacy, and despite the busy research schedule, I found time to write some long letters. Often, especially when I was away from home, I used to write a letter or postcard to my children. Mimi wrote back more often than the other two, but even her letters came at long and irregular intervals. It was different with Margareta Claesson, a young Swedish girl I knew from my many visits to Stockholm. We had corresponded for several years, ever since I had sent her a postcard from Viet Nam when she was fifteen. She had replied with a long letter, and we continued writing. Over the years our letters became more frequent and more personal. Margareta told me about her school, her friends, dances and vacations, and the joys and sorrows in her life. Although I was rarely in Stockholm, we became good friends. One evening when I sat in my cabin, I started writing to Margareta about a conversation I had had late the night before with one of the crew.

After a long day in the lab, I had gone to the mess, where I found an off-duty member of the crew. As I walked in, he said, "207 days."

"What do you mean? To what?" I asked.

"To San Diego Day, the day we are in home port! You can't believe how glad I will be when this cruise is over," he continued. Anchored on the Amazon, his only wish was that 207 days of his life would vanish so he could be home again. The next time the ship called at Manaus to take on a new team of scientists, he would get drunk, pay for a girl, and have a break from his tedious routine on the ship.

It was past midnight, and we sat talking for a long time. "What I am really looking forward to," he explained, "is to be home for Christmas. My four-year-old daughter wants a giant doll more than anything else in the world, and when we go through the Canal, I will buy the biggest doll I can find in all of Panama."

Our talk drifted to the previous Christmas. "The ship returned to home port a few days before Christmas," he said. "I was waiting for nothing else during the long months on board, but after a couple of days the wife started nagging, said I was in her way, so I went out with the boys and got drunk. Christmas was coming, and that was what I was really looking forward to. And Christmas Day was good, but I couldn't sit around all day and do nothing. The wife got impatient and the kids started bickering about who got the biggest gifts. So I went out to see if some of the boys were around. Got a little drunk, perhaps, but what else was there to do?" He explained that by New Year's he could hardly wait for the ship to sail again, and in January he was glad to be back on board. Then it started over: first a few days of relief from the tedium at home, then counting days and dreaming about the future; stopping in some port en route, getting drunk, paying for a girl; more boring months on board and crossing off the days until San Diego and a wonderful Christmas with the kids.

"How will it be this year?" I asked. Certainly not like last; he was emphatic about that. He would have a good Christmas, he wouldn't complain about being bored, and he wouldn't count days until leaving port again.

For how long will that last, I thought. I wondered about this sailor, imagining that when his time comes and he knows he is about to die, he shouts: "Please, Lord, let me live again and next time I will do better, I will. . . ."

This sailor brought home to me something important—that it is today we live, not in some imagined future when we are going to be happy. We live now, not yesterday and not tomorrow. Tomorrow never comes; tomorrow always stays one day ahead of us. I said good night to the sailor and went to bed, thinking about a man who spent his life crossing off days, wishing they would disappear.

M Y TIME ON the *Alpha Helix* was stimulating and exciting. I was with a team of first-class scientists who were conducting a variety of interesting experiments. Yet my own plans faltered. My equipment was never found, so I couldn't carry out the studies I had planned.

My greatest disappointment was that we obtained only two sting rays. Our first sting ray died in an aquarium for no apparent reason. The second was caught on a hook and badly damaged, but before it died I obtained a blood sample. Its urea concentration was nearly zero, as low as in all freshwater fish. This was exciting because all other sharks and rays need a high concentration of urea in their blood, and it now appeared that the freshwater ray was unique in being able to dispense with a substance essential for all its relatives. As I had only the one sample from a dying animal, the result was never published.

With my own plans frustrated, I asked Kjell Johansen if I could help his group in their studies of air-breathing lungfish and electric eels. Kjell went about all his work with immense energy and almost compulsive determination. He was the only person I might find in full swing long before breakfast at seven, when I had things to look after. I spent my mornings keeping a log of the activities of the preceding day, looking over the animal supply, and trying to foresee the needs of the coming day, hoping to avert conflicts about the use of our animals, and be ready for the morning conference at eight.

The work with Kjell brought me in contact with research areas in which I was interested but had little direct experience. He wanted to know which factors cause a lungfish to rise to the water surface to breathe and fill its lung with fresh air. As their name implies, lungfish breathe with lungs rather than gills. The question is whether a lungfish takes its next breath when the lung air is depleted of oxygen, or when the carbon dioxide in the blood, a waste product of metabolism, accumulates and must be eliminated.

In air-breathing animals, like ourselves, an increase in carbon dioxide in the blood stimulates the respiratory movements. In contrast, in aquatic animals, such as ordinary fish, a lack of oxygen stimulates increased breathing. Would a lungfish respond like an air breather or like a fish?

To find out, we replaced the air above the lungfish aquarium with air containing less oxygen. We found that in the presence of reduced oxygen, the fish came to the surface more often to breathe. When we replaced the ordinary air with pure oxygen, the fish breathed less frequently. The answer was clear: lack of oxygen determines breathing rate.

These experiments said nothing about any effect of carbon dioxide. We knew that mammals increase their breathing severalfold if 5 percent carbon dioxide is added to the air. We found that the lungfish did not respond at all to increased carbon dioxide in the air. In this way we showed conclusively that breathing in lungfish is far more fishlike than mammal-like.

We also studied the air-breathing electric eel. At intervals of a few minutes an electric eel glides to the surface and takes air into its mouth, holding it there while it slowly sinks back. Inside the mouth are numerous protruding papillae, richly supplied with blood, which take up oxygen. Its breathing is fishlike; it responds to lack of oxygen and not to increased carbon dioxide, just like the lungfish.

The collaboration with Kjell made me feel like a student again. I admired his surgical competence and his excellent skills as a team leader. I felt privileged to be included in his research, and by the time the *Alpha Helix* returned to Manaus to pick up the next team, I could honestly say that I did not regret any part of the cruise, not even the loss of my equipment.

Finding a Friend

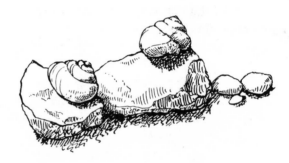

AFTER THE WORK in Brazil was finished in early July, I planned to go to Scandinavia. I had invited Mimi and my Swedish friend Margareta to take a car trip with me to Norway in a new Volvo I would pick up at the factory in Sweden. Mimi and Margareta were about the same age and knew each other well; most recently they had spent time together when Margareta was in Denmark to work with Mimi's cousin Bodil Krogh, a gifted textile artist. Later in July I would meet my friend Agi Bene in Copenhagen, where she would attend the International Congress of Psychoanalysis.

From Brazil I flew to Denmark, where I visited Niels and spent a couple of days with him and Ellen. There I learned that Mimi had returned to the United States sooner than planned; this disappointed me because I had really been looking forward to showing her and Margareta the country where I had grown up.

I continued to Gothenburg, where my brother, Patrick, was affiliated with the university hospital. I picked up my new car and drove to Stockholm to see my sister, Astrid. I had no other plans, and the weather was beautiful, so I suggested to Margareta that even if Mimi couldn't come with us, we take the trip to Norway just to try out my new car. She wanted to discuss with me her education and career plans and promptly agreed, and without further preparation we set off on an informal trip, taking each day as it came. Although she was twenty, I still thought of her as a teenage schoolgirl; otherwise I wouldn't have asked her to travel alone with me.

Margareta also looked like a schoolgirl, small and red-haired, a little chubby, and outgoing, with a wonderful, engaging smile. We enjoyed each other's company and had endless things to talk about. We had never spent much time together, but our long letters had led to a warm friendship, and we had a pretty good understanding of each other. I found her spontaneity and open mind wonderfully refreshing.

I liked to listen to Margareta tell me about her further education and choice of career. She had spent the past year in art school, and I considered her remarkably talented with a prolific artistic imagination. But instead of encouraging her versatility, art school had stifled her with its rigidly scheduled exercises. By the end of the school year she had become discouraged and decided to give up art and instead apply to teachers' college.

Margareta had a wonderfully curious mind, and it was fun to share my knowledge with her. I told her why the meltwater from glaciers has an opaque gray-green color, how cheese is made from goat's milk, and how plants change above the timberline, showing her the tiny leaves of the dwarf birch, foot-high bushes with leaves no bigger than mouse ears. I told her about quaking aspen, explaining that the leaves have flattened stems so that the slightest air movement makes them quiver. It was the Swedish botanist Carl Linnaeus who gave the aspen its scientific name, *Populus tremulus,* the trembling poplar. When we stopped at the roadside, Margareta picked wild strawberries, threaded them on a grass straw, and handed me a fragrant beaded string.

One evening we drove up a steep, rocky mountain road and spent the night in a timbered farmhouse several hundred years old. Our beds seemed equally old, hard and lumpy and too short for me to stretch out in. We laughed about the low ceiling, which kept me from standing up straight. The ancient glass in the tiny window distorted the valley below us into a surreal dream. This was the perfect setting to tell the Norwegian folktales that I had grown up with, tales about trolls and treasures, about the small people who live underground, and about beautiful fairy maidens who bewitch young men.

We crossed the high mountain plateau on the way to the west coast. Winter storms had piled the snow into high drifts that remained in mid-July. We stopped the car and had great fun running hand in hand and barefoot in the snow. We didn't care about the few other tourists, who looked at us as if we were crazy.

On our way back to Stockholm we stayed a couple of days in Oslo. There we saw the Viking ships, excavated from thousand-year-old burial mounds, and admired the beauty of their lines. We also saw the *Kon Tiki* raft on which my old friend Thor Heyerdahl had crossed the Pacific, and I told Margareta all about him and his husky dog. At the Maritime Museum we ate a superb fish dinner, better than any we could remember.

One morning we walked down to the harbor in the center of the city. In front of the monumental city hall, a small trawler had just come in with fresh shrimp, cooked on board within minutes of being pulled out of the water. I bought a bagful of the warm shrimp, and we sat at the end of the pier, our feet dangling over the edge, eating shrimp and throwing the peels in the water.

With a colleague, Johan Steen, who had been on the *Alpha Helix* with me, we went to see Edvard Munch's paintings at the Munch Museum. Both Margareta and I were deeply moved by the remarkable power of Munch's paintings, which express so much of the inner turmoil and tortured mind of this great artist.

Enjoying the beautiful weather, we remained in Oslo before returning to Stockholm. Soon afterward Margareta was accepted at teachers' college and started living in the old university town Lund. These were difficult years for her as she tried to keep up as a normal student in spite of much illness and recurring problems in controlling her diabetes.

As agreed with Agi, I met her in late July in Copenhagen, where I enjoyed meeting her professional colleagues and friends. They were interesting people, ranging from the most formal professionals to strikingly informal extroverts. Agi introduced me to Anna Freud, a small, white-haired old lady, the daughter of Sigmund Freud and a distinguished psychoanalyst in her own right. Agi worked with Anna Freud at the Hampstead Clinic in London and had often spoken about her. I made a few polite remarks to Dr. Freud, but she didn't respond. It made me feel stupid, but Agi told me afterward that Anna Freud was totally inept at small talk and unable to carry on a casual conversation.

After a week in Copenhagen, we drove to Gothenburg to visit Patrick and then to Stockholm to see Astrid. My family and friends liked Agi. With her black, curly hair, dark eyes, and southern complexion, she looked exotic among the tall, blond Swedes.

Being with Agi was never dull. She had broad cultural interests, and we had more than enough to talk about. She was an intellectual and less doctrinaire about Freud and psychoanalysis than many of her colleagues. I especially liked that her vocabulary wasn't locked into a straitjacket of incomprehensible professional jargon. I suspect that the mumbo jumbo many professionals use, not only in psychoanalysis but in many fields of science, often serves as a cover for lack of clear thinking.

From Sweden we drove to Norway. In Oslo I wondered how Agi would react to Gustav Vigeland's monumental sculptures. Vigeland, who died in 1943 at the age of seventy-four, is much admired by many Norwegians and equally despised by others. Vigeland had an artistic dream that resulted in a unique contract: the city of Oslo granted him a studio and workshops at no cost, and in return his entire artistic production became the property of the city.

Vigeland's sculptures are located in a large park at Frogner. The centerpiece is a huge monolith cut from an enormous block of granite. A massive fountain is surrounded by large groups of sculptures, and a stone bridge that crosses a small lake carries hundreds of human figures. As a student I had admired some of Vigeland's early sculptures, especially a stunning monument to one of the world's great mathematical geniuses, Niels Henrik Abel, who died in 1829 at the age of twenty-seven. But as I grew older I more and more disliked Vigeland's oversized figures; every one expresses a Teutonic quality that seems to lack human feeling and tenderness.

Agi found the massive display repulsive. She had grown up in Budapest and lived there through the war. She had found the body of her father, murdered by German occupation troops; had survived the invasion by Russian forces; and after the war had gone to Switzerland and then England. After learning of the horrors she had lived through, I knew how fortunate I was to have spent the war in relative safety in occupied Denmark.

We parted in Gothenburg, where I shipped my new car to the United States while Agi flew back to London.

Back at Duke I resumed my usual life. My research and my name had become well known among biologists, both in the United States and elsewhere. I became chairman of the National Advisory Board for the *Alpha Helix* at the Scripps Institution and worked on a variety of national committees. Although I still traveled a great deal, I spent most of my time in Durham, attending to my teaching and my active research program. I was surrounded by an international group of dynamic and stimulating students and young postdoctoral associates.

I spent long hours in my office attending to correspondence, grant applications, and reports; writing scientific papers; and revising my co-workers' drafts. Every year I revised my lecture notes or wrote new lectures. I took teaching seriously; even when I was busy with other matters I never went unprepared to class.

I liked living in my new house, less than ten minutes from Duke by bicycle. I woke early, usually about six, and was normally at work before anybody else. The greatest advantage of living close to Duke was that I could go home for lunch and even take the time to listen to a record or read for a while before returning to work. Lunch was my first meal of the day; in the morning I drank cup after cup of strong tea with milk, like most people in England. When my secretary and the others left at five, I usually stayed for another hour or two of undisturbed work before going home for dinner. After dinner I usually went back to work. When I stayed at home, I listened to records, read, and wrote letters.

Whenever I had visitors, often from overseas, I invited them to stay with me and gave parties for them, and on those occasions my evenings were lively and stimulating. Being alone most of the time was something I never liked and never got used to. I wasn't eager to be married again, but I felt a need for a close and trusted friend.

Shortly after I moved into the house, Mimi stayed with me briefly. She liked my house, the open spaces, the simple furniture, and the large windows with a view of big trees and untended shrubs. She said it was just the kind of house she wanted me to live in. The year before, I had visited her in Massachusetts, where she finished high school. Now in Durham, she was no longer merely a schoolgirl, but a soft-spoken, engaging young woman with long blond hair and a wide, warm smile. We both felt comfortable with the change, and I felt that we understood each other.

During the next two years I saw Mimi only occasionally. In 1967 she left for Reed College, in Portland, Oregon, where I visited her briefly after lecturing in Seattle. She had many good friends, and I spent a long evening and half the night with an interesting group of intelligent young people, discussing their concerns about the society we lived in. During her Christmas vacation Mimi stayed with me in Durham, and we continued to enjoy each other's company.

After a year at Reed, Mimi went to Denmark for the summer, where she worked with the textile designer Bodil Krogh. Mimi's artistic talent found a welcome expression in silk prints, but she wasn't happy. Over the years, the epileptic seizures that had begun when she was about seven had increased in severity. A physician had prescribed a change in her medication, but the alternative had disturbing side effects.

During the next few years I heard less from Mimi. She left college, moved to southern California, and occasionally sent me a brief note, but never said much about herself or what she was doing. She struggled hard to be independent and didn't ask for help.

Bent had lived with me while working as Dan Tosteson's research assistant and proved the most pleasant housemate one can imagine, but after he moved to California to work with Pete Scholander I saw little of him.

I saw more of Astrid. She lived in Washington, and when I was there for meetings, I always stayed with her. Since Astrid's in-laws also lived in Durham, she and her family often came for a few days or a long weekend. Almost every year I spent Christmas in Washington with Astrid, and afterward they came to Durham for New Year's, when I always had a big party for friends and colleagues.

It was a relief that I never saw Bodil. In 1964 she had moved to Cleveland as professor of biology at Case Western Reserve University, and I saw her only at a distance when we happened to attend the same scientific meetings.

During this period I spent some time with a pleasant female companion; we went on trips together and occasionally visited her parents or her brother, but I never wanted our relationship to become permanent. Our friendship filled some of my needs for companionship. Overall, however, for long periods I was alone, worked hard, and didn't find much satisfaction in anything else.

In July 1968 I was in Jerusalem to give the opening lecture at a meeting of the International Union of Pure and Applied Biophysics. Though I am certainly not a biophysicist, neither pure nor applied, I was invited to speak by Richard Keynes, a distinguished English physiologist known for his studies of nerve impulses in the giant axons of squid and of high-voltage discharges by electric eels, a phenomenon related to the movement of ions across cell membranes.

The biophysical meeting dealt with these kinds of problems—how water and dissolved substances cross cell membranes, how nutrients are absorbed from the intestine, how the salivary glands make saliva, and how urine is formed in the tubules of the kidney. I had been interested in kangaroo rats, saltwater frogs, and salt glands of marine birds, but I had never been involved in research at the level of the cell membrane.

The speakers at the meeting were required to submit abstracts of their talks for advance distribution. I looked over the 112 abstracts I received and realized that a talk on desert animals would not be very relevant to the world's foremost specialists on cellular transport in intestine and kidney, red blood cells, frog skin, and toad bladders and in the function of artificial membranes. I felt like a rank amateur among highly sophisticated investigators in a field in which I had little experience.

Then it dawned on me that all these high-powered abstracts dealt with membranes at a liquid-liquid interface, that is, membranes with liquid on both sides. For example, a red blood cell has a membrane that separates its contents from the surrounding blood plasma. Cells in the kidney separate the cell contents from the liquid urine. In a fish gill the cell membrane separates the cells from the surrounding water, as do the skin cells of a frog when it swims in water.

But if the frog sits on land, the cells at the surface of the skin have one side exposed to air. Every land animal has at its outer surface a liquid-air interface, which presents a major challenge: How is water loss regulated when the surface cells are exposed to air? None of the 112 abstracts dealt with the important problems of the liquid-air interface. Here was something I knew more about. I called my lecture "The Neglected Interface," the title intentionally cryptic to stimulate curiosity.

I began by discussing the need for desert animals to conserve water. I described how the kangaroo rat reduces evaporation from the respiratory tract by exhaling air at below body temperature. I then discussed how difficult it is for moist-skinned animals to reduce evaporation from the skin. Water evaporates from the moist skin of frogs and snails at about the same rate as from a free water surface, so these animals must avoid excessive water loss by their behavior. Many are much more tolerant of dehydration than mammals and can rehydrate rapidly by uptake of available water through the skin.

Reptiles, such as snakes, lizards, and turtles, have dry, scaly skin, and were long thought to lose virtually no water through the exterior surface. However, work by Peter Bentley in my laboratory showed that reptiles lose a great deal of water through the skin. A desert tortoise or a desert lizard, however, loses only one twentieth as much water from each square centimeter of skin as does a water snake or an alligator. Thus desert reptiles are better able to conserve water than their relatives that have water more readily available.

As for insects, more than a million species live in every imaginable terrestrial environment, except polar ice caps and glaciers. Numerous insects live in the driest deserts, using a mechanism that at first glance seems impossible: the ability to absorb water directly from the surrounding air.

I described how a desert roach, for example, would lose water if held in very dry air. But if placed in moister air—say 80 percent relative humidity—the roach would take up water to cover the deficit. How insects manage this process is indeed remarkable, for at 80 percent humidity other animals, even desert lizards, lose water by evaporation. Even more remarkably, the desert roach would take up water until it had regained its original weight and water content, and then cease further uptake. In other words, the uptake was a well-regulated process, serving only to make up the deficit. I hoped to challenge some interested biophysicist in the audience to help solve the riddle of how these animals manage to pull water molecules out of the air.

After finishing my own lecture, I faithfully sat through the lectures of others to learn about membrane transport. However, I often failed to understand the esoteric equations that some speakers seemed to consider more important than what happens in living animals. After two days I lost patience and went to look at the deserts with my colleague Arieh Borut, who had worked with me on salt glands.

In the cool early morning we set off to Hebron, the birthplace of Abraham. As the heat increased, we continued to the Dead Sea and to Qumran, where the Dead Sea Scrolls were found in a cave. We then took a hot and dusty road to Jericho, where according to the Bible, the walls crumbled when the Jews marched around the town, blowing rams' horns.

This biblical place is now an enormous mound of stone and rubble with a deep exploratory archaeological trench cut from the perimeter to the middle of the mound. We climbed the mound and stood at the edge of the trench, looking down to the bottom, perhaps fifteen meters below, where a round tower was partly uncovered; Arieh said that it dated from the Stone Age. The sides of the trench revealed perhaps thirty layers of accumulated debris, each separated from the next by a band of darker color, and each representing a separate cultural epoch.

As I turned away from the trench I was surprised to see standing behind me Hans Ussing, my old friend from my student days in Copenhagen. Ussing, one of the world's most distinguished investigators of membrane transport, had, like me, left the meeting to see some of the wonders of this part of the world, although our reasons for leaving the meeting were probably quite different; I understood too little of the lectures, and he already knew too much. We greeted each other cordially and exchanged a few words about the trench. If the lowest layer was 6,000 years old—a reasonable guess—then each layer represented on the average a culture that had spanned a couple of hundred years. It was humbling to consider our own minuscule significance in the immense span of human history.

My LECTURE in Jerusalem reminded me of an interesting problem I had encountered a couple of years earlier during a visit to Israel. I had been walking in the Negev Desert with Amiram Shkolnik when I saw, among the small rocks and pebbles, the small shell of a dead snail, bleached and chalky white. "How did this shell end up here in the desert?" I asked.

Amiram's reply astonished me. "It isn't dead," he said. Then he explained that in the hot desert summer the snails withdraw into the shell and close the opening with a thin, chalky membrane, an epiphragm. Most of the year they remain dormant and inactive, fully exposed to sun and heat, coming out and feeding only after rains. However, in the desert it rains only once or a few times each winter, and sometimes not even once.

How can the snails survive the heat, and how do they manage so long without drying out, I wondered. With a study grant from the National Geographic Society, I went back to Israel in the summer of 1969, taking Dick Taylor with me. Amiram helped us settle in at a field school in the Negev, where the three of us could live and work.

The snails posed three obvious puzzles. First, how do they avoid dying from presumably lethal temperatures? Second, how do they avoid desiccation when there is no source of water for a year or more? And finally, why don't they starve to death when eating nothing for so long?

A small pebble the size of a snail shell sitting in the midday sun on the desert surface heats up to some 60–65°C (about 140–150°F). To determine if a dormant snail could tolerate such temperatures, we had to establish their lethal temperature. If we exposed snails in the laboratory to 60°C, they all died within minutes. At 55°C half were dead within two hours and all of them in eight hours, but at 50°C (122°F) all the snails survived for eight hours, roughly the duration of a day in the desert. Such resistance seemed reasonable, for 50°C is about as high as any animal can tolerate. So far, the snails did not seem uniquely tolerant of excessive temperatures.

Did the snails have a way of keeping a lower temperature than the surrounding rocks and pebbles? To answer the question we drilled a tiny hole in the shell, inserted a fine thermocouple to measure the temperature, and put the snail in the sun. To our amazement the snail did not get as warm as a small pebble lying next to it. Measurements showed that the chalky white shell reflects much of the solar heat, both visible light and the radiation in the infrared part of the spectrum.

In addition, the snail withdraws to the smaller top whorls of the shell, so that the air in the largest whorl serves as an insulating cushion against the heat from the ground surface below. The temperature in the elevated part of the shell, where the snail sits, remains well below the snail's tolerance limit of 50°C.

We next examined the snail's ability to remain alive in the absence of water by determining how fast a dormant snail loses water. After weighing dormant snails and then leaving them undisturbed on the desert surface for several weeks, we reweighed them and found out that the average daily water loss was less than one milligram per day, perhaps as little as 0.5 milligram.

A full-sized snail contains about 1.4 grams, or 1,400 milligrams, of water. If a snail can survive the loss of half of its body water, or 700 milligrams, it should be able to go without water for some four years. Thus, the snail's water reserves will last for several years, even if the winter rains fail for a year or two.

To answer our third question, how the snail manages a year or more without eating, we needed to determine the metabolic rate of the dormant snails. This measurement required more sophisticated equipment than we had in the desert, so we carried dormant snails back to Duke. The amount of oxygen the snails consumed was, for long periods, so low that we were unable to measure it accurately. The snails had virtually suspended their metabolism during dormancy; however, at intervals of several days their oxygen consumption increased to measurable levels. We could then calculate the rate at which the snail might metabolize its stored nutrients. If a snail consumed an average of 5 microliters of oxygen per hour (much higher than the average we measured), it would use approximately half of its nutrients in forty-eight months, or four years. Thus, the dormant snails do not risk dying of starvation even if the winter rains fail and they eat nothing for a year or two.

What is there to eat in the barren desert? When it finally rains, the snail comes out and creeps over the rocks and pebbles. It scrapes off the softened surface layer of the rock with its sharp tongue, ingesting and digesting the microscopic algae that live just below the surface of the limestone.

DURING THIS time Margareta wrote frequently from Lund, telling me about her personal life, her studies, the difficulties in controlling her diabetes, and a bleeding in one eye that warned of the possibility or even likelihood of blindness. I replied with equally long letters, often setting down my thoughts but saying little about my personal life.

The slow death of an aunt in a hospital for the chronically ill prompted me to write about illness and death and what seemed to me the immorality of keeping human bodies alive when they are no longer human beings.

I wrote, "In my teaching laboratory we can keep an isolated turtle heart alive and beating for hours or even days. More sophisticated medical techniques are now so advanced that, not only can human organs be kept alive and perfused for long periods of time; we can keep a whole human body functioning with the aid of respirators and other mechanical devices. Where then is the boundary between life and death? We must let humans die as humans, not as laboratory preparations. We must use our medical resources to improve life for those who are still alive and can be helped.

"As we grow older, we think more about these problems. Not so much that I am afraid of growing old, but I am not looking forward to years of gradually declining body and mind. A scientist friend of mine was not long ago found dead in his laboratory, 'due to self-inflicted poison' the notice in *Science* said. He suffered from incurable cancer, and I think of the long and painful misery he spared himself and his family."

A year later, in 1968, on a long trip that included a visit to Stockholm, I found Margareta changed to a young woman. On returning home I felt alone in my house. Materially I had all I could wish for. I liked my work and felt I did it well. I had young people around me and knew I could be of help to them. But I also felt more unsettled than before.

In a letter to Margareta I described my irregular eating habits, which wouldn't suit her because she needed regular meals to keep her diabetes under control. I asked if she would like to visit me. We both toyed with the idea, but I was ambivalent and in October 1968 I wrote: "You would like it here. There is so much I want to show you and tell you. But the more I think about how much pleasure and joy we could have, the better I understand that you are no longer the young girl that you were when we traveled together in Norway. Then you were a schoolgirl; now you are a woman. You are terribly attractive, and if you came here, our relationship likely would never again be the same as it is now. Perhaps it shouldn't matter much, but it nevertheless is very important for us both."

Margareta wrote about a friend in Lund who had become pregnant and gone through the agony of deciding what to do. I was greatly relieved when Margareta wrote that her friend had decided in favor of abortion, and I wrote about my thoughts. "For me it is difficult to understand that society is so fiercely opposed to abortion—it seems that the purpose is to punish a woman for what she has done while a man goes free. What will it cost in future suffering for the woman and for a child not yet born who never had anything to say about it? The righteous men and women who set themselves up as the guardians of our society's morality can feel smug, for they did not cause the suffering."

About this time I briefly joined a group called Parents Without Partners. The divorced wife of a faculty colleague asked me to accompany her to a party. Most of the people there were divorced; a few were widowed. It was lively, with some dancing and drinking, but I found the party a depressing experience. Some drank too much. Others were obviously on the lookout for sex. Some sat quietly, looking depressed and miserable.

In one letter I told Margareta that I could not continue seeing Agi as before, for one cannot continue an emotionally demanding relationship on an on-again off-again basis. Since the summer of 1967 I had known that I was unable to make our relationship last. At that time Agi had come to Durham for a few days, but I was uncomfortable having her in my house. We understood each other too well, yet we were too different. What brought us together and kept us friends was the deep understanding that is possible when both have been through an intensive analysis. But our insight also caused difficulties.

Margareta asked if I had known Agi when I was still married to Bodil. I replied that I met Agi much later, in London, that no woman was involved in my separation from Bodil, and that I think playing around is terribly destructive to a marriage.

I also told Margareta about my children. Astrid seemed content but was planning to continue her education when her children grew older. Bent had left MIT and was working in New York as stage manager for a small experimental theater. As with nearly everything he tried, he was very good at this job. When something went wrong, he had the ability to identify the trouble instantly and fix it. I knew less about Mimi. She lived in California with a boyfriend and managed, at least periodically, by working as a cocktail waitress. "What can one do for one's children when they are over twenty, except wait for them to grow up?" I asked.

In one of her letters Margareta expressed concern that she had been a nuisance on our travels in Norway, so I replied that she had been the easiest and most uncomplicated travel companion I ever had—never any complaints, never any bother, always willing and eager for what we were doing and seeing. Being with her had been wonderful because she had simply enjoyed it all.

Still, in my letters to Margareta I usually didn't mention female company, and it must have shocked her when I wrote in 1970 that I was planning to get married by early summer. I would rather not think about this unfortunate episode, for within weeks it was obvious that the marriage was a catastrophic mistake and that any attempt at continuation would be futile. With the help of lawyers we arranged for a separation and later a divorce. Under the agreement my divorced wife would remain in the house for a year; thereafter I would continue supporting her and her teenage son.

I moved to an apartment and resumed a monastic life of hard work and long hours in my office. I saw few friends. Dick Taylor had moved to Harvard, and I missed him. I hated myself for having behaved like a fool. I knew that I could handle science better than my personal life and concentrated on work.

I postponed writing to Margareta about the divorce because I didn't want to complain about my difficulties. When I finally put my thoughts on paper, I confessed that "I feel like an old and disappointed man. I keep to myself and avoid people and friends."

I had more work than I could handle, which gave me little time to think. I was interviewed by the BBC, gave lectures at other universities, and stopped in Washington to see Astrid as often as possible. I struggled with being alone again. I was depressed, but managed to keep up with my work.

At the end of November I decided to spend Christmas in Stockholm with my sister and began to look forward to a few days of skiing or ice skating. I left for Stockholm just before Christmas and remained there until the beginning of January. The peaceful Christmas with my sister and her family was just what I needed. I enjoyed her quiet, friendly home and the familiar holiday food I had grown up with in Norway. Besides, I could spend time with Margareta.

The many lakes around Stockholm were covered with wonderful thick, smooth ice, excellent for skating. One day Margareta and I went skating with her parents on a large inland lake, gliding over the ice for mile after mile. On our way back, Margareta and I sat in the back of the car. Her feet were miserably cold, and she removed her boots and sat with her feet in my lap while I tried to warm them between my hands. I was worried because inadequate circulation to the feet is a serious complication of diabetes and may lead to gangrene and amputation of the legs. Her feet stayed cold, so I put them under my clothes, directly against the warm skin of my stomach. We sat like this until we were home again, understanding each other and not needing to say a word.

Before New Year's, Margareta took the train to Copenhagen to spend a few days with a friend who had come from London to see her. She was now a grown woman, leading her own life. I drove her to the train station, and when she had left I felt old and disillusioned, bereft of my young, warm friend. The spark was gone from my last few days in Stockholm. Much later Margareta told me that, had I asked her to remain in Stockholm, she would have done so.

THE FIRST WEEK after my return felt terrible, but the long hours of work and my good friends in the laboratory kept me too busy for much introspection. By summer I would be back in my own house, and in the meantime I had numerous obligations to occupy my time.

Two weeks after I returned I was in Washington for a committee meeting to award travel grants to scientists wanting to attend the International Physiological Congress in Munich that summer. The committee had received 260 applications, but the funds would cover only about 100 grants. Before the meeting each committee member had spent at least sixty hours reading all applications.

When we met in the magnificent marble building of the National Academy of Sciences, it was fairly easy to pinpoint the top 50 or 60 applicants; I was amazed at how readily we agreed, even though the committee members represented widely different fields of specialization. Likewise, we almost as readily agreed on 50 or perhaps as many as 100 that were low on the list. This left us with a group of 100 or more applicants in the gray zone; each had to be scrutinized, discussed, and evaluated. We sat in meetings for two full days to reach a presumably equitable decision after again going over every application.

This work is typical of what scientists do as volunteers, taking time from other work. None of the major granting agencies, the National Institutes of Health or the National Science Foundation, could operate as they do without the immense amount of volunteer work done by scientists on review panels and committees.

LATE THAT SPRING I would be at Cambridge University. The zoologist Torkel Weis-Fogh had written that I had been chosen to serve as the Hans Gadow Lecturer for the academic year 1970–71 and was to lecture toward the end of the Easter term. It was not only the honor of serving as Gadow Lecturer I was looking forward to; I would also see more of Torkel, whose stimulating company I always enjoyed. I had visited him several times in Copenhagen, where he invariably invited me home for dinner with his wife, Hanne. After they moved to Cambridge in 1966, I continued to visit him whenever I could.

I felt challenged to be lecturing at one of the world's greatest universities, and to be associated with what I thought of as *the* outstanding zoology department in the world. My lectures had to be as good as I could possibly make them, good enough to satisfy those who had invited me.

I set aside time the same way I had a decade before when I wrote the monograph on desert animals, scheduling one full week for the preparation of each lecture. I had already gone over in my mind what I wanted to discuss and devoted every minute to nothing but the preparation of my notes. I usually lecture with only brief notes, but this time I wanted to have a complete manuscript for each lecture so that I would not risk getting lost in the middle of one. I would also give some informal seminars, but for these I trusted my usual routine.

After I had the complete typescript for my lectures, I prepared my slides. I needed many new diagrams, which I drew because it was easier to do them myself than to explain the work to a professional draftsman. The photographic work I turned over to the technician in my department, who always finished my slides by the next day. With his help I had everything ready well before my scheduled departure.

Three days before I was to leave, I received a shocking cable from Arthur Ramsay, the acting head of the department at Cambridge: "REGRET INFORM YOU TORKEL AND HANNE IN CAR CRASH STOP HANNE DEAD TORKEL RECOVERING NYBORG STOP LECTURE ARRANGEMENTS UNCHANGED STOP."

After that brief message I received no word about the accident or Torkel's condition before I left. On the plane from New York to London, the tragedy again overwhelmed me.

The master of Christ's College, the distinguished chemist Lord Todd, had invited me to stay as a guest of the college while I was in Cambridge. What I knew about college life came mostly from C. P. Snow's novels about internal politicking and backbiting, matters I was unlikely to become involved in. When the taxi from the train station pulled up in front of the beautiful low building of Christ's College, I entered the imposing gate, whereupon I saw a discreet sign on a door to the left that said "Porter's Lodge."

As I entered with my suitcase, a gray-haired man in a black suit behind the counter politely inquired, "I assume you are Professor Schmidt-Nielsen?" Then he handed me a pile of letters and said that he would call Dr. Coombe. I recognized Margareta's handwriting on a letter with Swedish stamps; the others were dinner invitations from colleagues, Arthur Ramsay, Hans Lissmann, and several more.

Moments later Dr. Coombe came and introduced himself. Of middle height, slim, with brown hair, he didn't look like a formidable intellectual who might make me feel an unworthy intruder. He put me at ease by talking about the weather and the drought—there had been no rain for three days. He was a botanist, he said, and as he was a bachelor and lived in college, he had been assigned the pleasant task of taking care of me.

Dr. Coombe gave me the first details about the accident. Torkel and Hanne had been in Denmark and were on their way back to England in their new Volvo when an oncoming car veered into their lane and hit them head-on at high speed. Hanne died from a ruptured liver and internal bleeding. Torkel's chest was crushed; he was in critical condition but was expected to live.

After lunch Dr. Coombe took me to my rooms, an apartment belonging to a chemist who was in North America on sabbatical leave. We entered a large corner room with two large windows, huge bookshelves, and a small gas heater where the fireplace should normally be, surrounded by beautiful blue and white tiles.

In the enormous bookshelves, a second row of books sat behind those in front. "How can he read the titles of the books in the back row?" I wondered. There were approximately 130 shelf feet of books, and if they were worth $20 per inch (cheap for scholarly books), the total would be worth more than $30,000. Later I learned that the owner was a Syndic of Cambridge University Press (the "Syndicate") and therefore received a copy of every book the press published.

I ate my dinner in the dining hall, where the fellows and their guests assembled in an anteroom before they filed in while the students stood at their places. We found our places at high table but remained standing until a gong sounded, the signal for a student to say grace in Latin. Then the students were allowed to sit and enjoy their food, which I gathered wasn't identical with that served at high table, where wine was served every day.

From my place at high table, I had a good view of a portrait of Charles Darwin, once a fellow of Christ's. After dinner, at the port served in the fellows' room, Coombe showed me an old record book where one of the entries read: "Mr. Darwin too late at high table, fine 1 bottle." Another entry recorded a bet that was penned in such cryptic terms that today no one could decipher it, but the outcome was clear: Mr. Darwin lost and paid one bottle. A good way for the other fellows to get free after-dinner port or claret, I thought.

On my first Sunday at Cambridge I went to the cathedral-like King's College Chapel to listen to the music. The organ and the magnificent boys' choir were remarkable. Most of the music I recognized as Handel's, one of my favorites. In this setting it seemed more beautiful than any I had heard before.

One day shortly after I started lecturing, I met with the science editor of Cambridge University Press, Dr. Alan Winter, who agreed to publish my Cambridge lectures as a book. I emphasized that I wanted the book priced as low as possible, and when it was published in 1972 the hardback sold on the British market for £2.20 and the paperback for £.60. The corresponding U.S. prices were $5.95 and $1.95, so that almost any student could afford to own the book.

The title I chose, *How Animals Work,* was a bit of a pun. Physiology is about how animal organs function or work, and I also discussed the work of running and flying and swimming. My lectures had dealt with panting dogs and bird respiration, kidney function, whale flippers, the swim bladder of fish, animal locomotion, and a variety of other subjects.

By the time the book was ready to go to press, Torkel had recovered sufficiently to be back at Cambridge, and he wrote the kindest foreword anyone could wish for, saying that "although firmly rooted in classical physiology and morphology, Schmidt-Nielsen is never conventional and is able to see new and surprising solutions and relationships. It is the mixture of classical scholarship and common sense with the innovating spirit of an artist which has led him and his collaborators to an understanding of how animals survive in deserts, the existence and function of salt glands in birds and reptiles and, as it now seems, to the solution of a classical problem, the function of the bird's lung."

I could not have wished for finer words from Torkel, a scientist I truly admired. This little book ultimately became quite cosmopolitan; in addition to English, it was published in Russian, Japanese, German, Spanish, and Greek.

When I had finished lecturing at Cambridge, I set out for Denmark to visit Torkel instead of going directly to Stockholm as I had intended. On the plane to Copenhagen I thought only about Torkel, Hanne, and the accident. Niels met me at the airport. He and Ellen lived in Birkerød, about twenty kilometers north of Copenhagen. On our way to his house his calm was a relaxing contrast to my distressing thoughts about Torkel.

At their house I put through a call to Torkel's brother, Jørgen, who was a hospital physician in Copenhagen. I had never met him, but I had called him from Cambridge to tell him I was coming to Denmark. He told me that Torkel was eager to see me, and he wanted to drive with me to Nyborg, where Torkel was hospitalized. It was a two- or three-hour drive, and on the way he told me more details about the accident. Torkel and Hanne had been driving west in a strong side wind. A car coming toward them at high speed veered into their lane as the side pressure of the wind was suddenly decreased in the wind-shadow of a house. There was no way to avoid the head-on crash.

Both Hanne and Torkel had been wearing seat belts and were conscious after the accident. They were quickly taken to the hospital by ambulance, but Hanne's internal injuries were so severe that she could not be saved. Torkel, whose rib cage was crushed, had had the presence of mind to instruct the ambulance personnel how to place him so that he could breathe with his diaphragm.

When we arrived at the hospital, Jørgen left me alone to see Torkel. I was amazed to find him sitting propped up. He later explained that he could breathe better sitting than lying down. "He looks almost normal," I thought. Though weak, he was lively and alert, and our conversation was as brisk and interesting as always.

Jørgen's account of the accident spared me from asking awkward questions. However, Torkel talked about the accident and about Hanne. He didn't seem to tire. He told me that one of the broken ribs had nearly punctured his heart. His condition was in fact still critical, but he had a strong will to be in control of his recovery. He was, as before, the realistic intellectual, ready to be in charge of his future. We talked for hours, and I left feeling great admiration for my friend.

On the way back to Copenhagen I spoke little. My thoughts were on Torkel and our long talk. When Jørgen dropped me off at Birkerød, I told him how deeply grateful I was that he had taken a full day from his hospital schedule to ease my visit with Torkel. The next morning I took the day train to Stockholm, where I arrived in time for a late dinner at Astrid's house. Margareta, who was teaching in a nearby public school, also came, but she had a bad cold, and it worried me that she complained about chest pains.

The next day Margareta's father lent me his car, and I took Margareta to the doctor, and then to the hospital for an X ray. It turned out that she had a low-grade pneumonia. On Sunday I went to her apartment to see her again, and we had a cup of tea while we talked. She was not only sick; she was also depressed. Her friend in London had told her that there could be no future relationship between them. She felt rejected and abandoned. Her illness would require weeks of sick leave that would keep her inactive. She also didn't know if she would be well enough by summer for a long trip by car with her good friend Gunilla to Italy and Vienna.

I told Margareta that I would be back in Europe to give a major lecture at the International Physiological Congress in Munich the last week of July. I asked her to consider stopping over in Munich on her way back to Sweden, although we both knew the prospect of meeting there was remote. Margareta was ill and overworked and might not be well enough to travel around Europe by car. When I left Stockholm on the night train to Copenhagen, Margareta remained in my thoughts. I was worried about her.

ON THE LAST day of May 1971, I moved back to my house to live again among my own things. Everything in the yard had been neglected for a year, and each morning I spent an hour or two outdoors, letting hard physical labor compensate for past frustrations and anger.

I didn't have a real garden, no flowers or grassy lawn that required regular mowing. When the house was built most of the surrounding land was covered with pines and deciduous trees, and I had planted more pines. They grew rapidly, and unless I thinned them the whole property would become overgrown and unmanageable.

I worked outdoors early in the morning, but still arrived at work before anybody else. I had only six weeks at home before I was to leave again, and I needed the time to prepare the lectures I would give that summer. First there was a symposium on desert biology in London, organized by Geoffrey Maloiy from Nairobi, who had asked me to talk about desert snails. That didn't require much preparation because I had lectured about them before and didn't need any new slides. Geoffrey had also asked me to summarize, at the end of the symposium, the major advances that had been discussed at the meeting. This talk obviously couldn't be prepared beforehand.

The lecture in Munich, however, required careful preparation. I was to talk about the studies Dick Taylor and I had carried out on the energy cost of animal locomotion. We had determined how much extra energy an animal uses when it runs, compared with sitting still, and we had accumulated this information for a variety of animals of widely different body sizes. In addition to giving an overview of our work, I wanted to compare the energy cost of running with other modes of locomotion, such as swimming and flying.

First I would discuss running animals. Dick and I had found that the cost for an animal to run a given distance is independent of the speed at which it runs. This notion may sound strange, for when an animal runs faster, it breathes harder and more. However, it arrives at the goal sooner, and overall, the energy the animal has used for the entire distance is the same. This general rule held for all the animals we studied, from mice to dogs. Other studies indicated that the same holds for human running (but not for walking, nor at sprint speeds, at which the runner doesn't reach a steady state of oxygen consumption).

Another interesting finding is that the larger an animal is, the less it costs to move one unit of body weight over one unit of distance. For example, it costs less to transport one gram of dog over one kilometer than it does one gram of mouse. Even animals as large as horses fit into this general scheme.

As for birds, we know that a flying bird must use energy to keep itself in the air and also that it encounters air resistance, or drag. For running animals, air resistance is a minor energy cost, but birds fly much faster than mammals of the same size can run, and for them drag is important. We can recognize this difference in the streamlined body of flying birds; for most mammals, drag is of minor consequence and streamlining is therefore unimportant.

The amount of energy needed for a bird to fly is not obvious, and measuring the oxygen consumption of flying birds is difficult. However, Vance Tucker, a colleague at Duke, had succeeded in training parakeets to fly in a wind tunnel, wearing a mask to collect exhaled air, and had made the first reliable measurements of birds in steady flight.

My students had measured the cost of flying for other birds and had found that the oxygen consumption of a flying bird is about the same as that of a mammal of the same size running close to its maximum speed. Yet it is far cheaper to move one gram of bird over one kilometer than it is to move one gram of mammal over the same distance. Why is this so? Because, although they have about the same metabolic rate while moving, the bird covers one kilometer in a much shorter time. When it comes to size, birds are like mammals: the larger the bird, the lower the cost of moving one unit of body weight over one unit of distance.

The overall conclusion is that flying is a more economical mode of transportation than running on the ground. This explains why a small migrating bird can fly nonstop for thousands of kilometers, using only its body fat for fuel. Consider how impossible it would be for a mouse of the same size to run without eating or drinking from a hypothetical wintering territory in South America to its summer breeding ground in Alaska.

Swimming is more complex. Because swimming animals are neutrally buoyant in water, they need not use energy to support themselves as do animals on land. But drag in water is much greater than in air and increases roughly with the square of speed. Although precise calculations are difficult, as a general rule we can say that for fish, the cost of swimming again relates to body size: a large fish uses relatively less energy than a small fish to move one gram of its body mass over one kilometer. In short, it is expensive to be small, whether you are a mammal or a bird or a fish.

For fun, I calculated what it would cost an organism as small as mammalian sperm to swim one kilometer. I used available data for the size of sperm, their swimming speed, and oxygen consumption. It takes 100,000 million bull sperm to make up one gram, roughly the size of the smallest fish for which we knew the cost of swimming. For this much sperm to swim one kilometer would require 10,000 times the energy a one-gram fish would use. Luckily for them, sperm need not swim one kilometer.

With my lecture written and the slides ready, I set out for London and the summer's first meeting.

DURING THE nearly two months since I had left Stockholm in May, Margareta and I had written many letters and often mentioned the possibility of meeting in Munich. She had written that the pain in her chest was finally gone; she was well enough to meet me in Munich, and I was delighted at the thought of seeing her.

At the hotel in Munich I registered and put my baggage in my room. I had my hand on the doorknob ready to leave when the telephone rang. I answered, surprised to hear Margareta's lovely, soft voice. She was in Kiel, far to the north, with friends. Could she come to Munich to see me, she asked. "Yes, of course," I said, and asked the hotel staff if they had a room for her. Surprisingly, despite the thousands of congress participants, they did.

Margareta arrived the day after my big lecture, and I arrived early at the station to meet her train. It pulled in, and there she was, smiling, in a summer dress, walking toward me with a tiny suitcase in her hand. I gave her a big hug. I had kissed her only once before, briefly, the preceding Christmas in Stockholm; it felt good to have my arms around her for a moment.

During the remainder of that congress I paid little attention to science and spent every minute I could with Margareta. We were happy together. We went to some lectures I wanted to hear, but otherwise we went to art galleries and talked and went for long walks. One day all congress participants had an evening meal in one of the large beer halls in Munich. Margareta and I sat at a table with Dick Taylor, Bill Bretz, and Don Jackson. She was delighted to meet some of my best friends, whom she knew from my letters.

Another evening there was a performance at the Munich opera, with the best seats reserved for congress participants. I had bought two tickets in the early morning of the first day of the congress, before any of the other participants had thought about tickets, so we sat in the front row and enjoyed *The Marriage of Figaro* with Fischer-Dieskau in the lead role.

On the last day of the congress I rented a car to drive to Göttingen, where we had a meeting on the respiration of birds. Bill Bretz and Don Jackson were going to the same meeting and came with Margareta and me in the car. A German colleague had recommended that we spend the night at an old castle, where they took in a few overnight guests. It was as romantic a place as anybody could wish, and Margareta and I went for a long evening walk in the nearby forest.

The meeting in Göttingen was only for invited participants, and it wasn't feasible for me to skip the lectures. Margareta therefore continued on to Stockholm while I paid full attention to science.

I returned home to Duke satisfied. I had given several good lectures, and my time with Margareta was unforgettable, leaving us even closer than before. We had been happy during the days we spent together, and we trusted each other, irrespective of what we could expect of the future. I missed Margareta, but I was unable to convey to her how deeply in love I was, and I did not understand how serious her feelings about me were and had been for years. We wrote to each other more than ever before. I told her about everything that happened in my daily life, not waiting for a reply before writing the next letter.

The fall was my teaching term, and I gave four lectures per week. I stayed busy, revising notes and reviewing new material. My colleague Vance Tucker suggested that we should have a computer in the laboratory, and I learned to write simple programs for it. It would be a full decade before IBM introduced personal computers and everybody became familiar with them; at the time, the thinking was that everybody would eventually be connected to big regional computers and that personal computers had no future.

We found our computer to be extremely valuable for the students. On a regional computer, every minute was recorded and eventually billed. But on our laboratory computer students could sit for hours and find out how the machine worked; they could make mistakes and take their time to solve problems without anybody looking over their shoulders. I too felt at ease not being connected to the mainframe computer. Before I started using our lab computer, it seemed formidable, but once I had learned to write simple programs, it was easy and saved me a tremendous amount of time.

That fall I was to write an article about bird respiration for *Scientific American* magazine. I needed a couple of complex drawings for illustrations, and the editor, Dennis Flanagan, sent one of the magazine's best artists, Tom Prentiss, down from New York to spend a couple of days working with me. I was impressed that expense was no concern. Prentiss was wonderful to work with, and keenly interested in animals and what we were finding out about bird respiration.

I thought a great deal about Margareta. We had several times lightly mentioned her visiting me during Christmas vacation, but we hadn't talked seriously about it. Then I made a resolute decision. On her birthday, October 14, I telephoned her. At that time transatlantic calls were a major, almost dramatic venture. It was evening in Stockholm, and her apartment was full of friends celebrating her twenty-fifth birthday. How good it was to hear her voice when she answered, warm and soft and totally surprised. I had only one question: would she come for Christmas? She sounded happy and at once said, "Yes, of course." I didn't know what more to say; I was overwhelmed, but there was no need for us to say more. As was so often the case, we didn't need words.

I wrote to Margareta about the arrival in New York, long lines to go through, immigration control, waits for baggage, customs control. But I told her I would meet her at the airport.

Before Christmas I drove to Washington, left my car at Astrid's house, and took the train to New York. I was at the airport long before Margareta's scheduled arrival. When I finally spotted her in the crowd coming out from the customs area, she smiled at me. It was wonderful to see her.

We took a train to Washington to see Astrid and her family. John and Astrid hadn't seen Margareta since the summer of 1963, when they had traveled with me in Norway and Margareta was seventeen. Then we set out for Durham in my car, finally alone. The air was crisp and cool, and the sun was shining from a cloudless sky. We had so much to talk about, and, more than anything else, we were happy to be near each other again.

On Christmas Eve we drove to the mountains in western North Carolina to visit John's parents. Gelolo McHugh, whom we called Mac, was a retired psychology professor from Duke; he and his wife had invited Astrid and her family as well as Margareta and me to celebrate the holiday with them.

The next morning both Margareta and I woke early, before daybreak. We went outside to look at the starry sky; in the darkness we could barely make out the outline of the valley below us. We stood looking out over the landscape while a line of deep red slowly spread along the eastern horizon and grew brighter. I stood behind Margareta and held my arms around her to keep her warm. We hardly spoke, only whispered a few words. We remained standing close together while the light slowly increased, until we heard others stirring in the house and went back inside.

When we returned to Durham we had a few quiet days before my big New Year's Eve party, an annual event. I cooked our dinners, and Margareta was always in the kitchen with me. Our meals were simple: some meat, lots of vegetables, a salad, and a glass of good red wine. I showed Margareta how to make yogurt and to bake bread, which doesn't take a lot of time or effort, only simple ingredients and common sense.

Early in December I had cut dormant branches of flowering quince from the garden, and at Christmas they were in full bloom, with pale pink flowers all over the leafless branches. Margareta was delighted to see delicate flowers in midwinter and wanted to draw or paint them. I had no paints, but gave her large sheets of paper, India ink, and brushes. She sat for hours at the large French windows, drawing flowers and looking out at the wintry garden.

Every evening we kept a big fire roaring in the fireplace. We talked for hours, stretched out on the rug and looking at the flames that slowly died down and left a heap of glowing embers. In the morning I used the remaining embers to start a fire with only a little kindling before putting on big logs to keep the fire going all day.

We got ready for our New Year's Eve party, expecting some thirty or forty of my friends to come. I didn't think of the house as very big, but we had room to house Astrid and John and their two children, Bent and his friend Maggie, and finally Peter and Karin Bentley from New York. We drank champagne and ate lots of good food. At midnight I made *glögg*, a Swedish drink of warm red wine and spices. I put the wine in a large bowl with a rack of sugar lumps over it, then poured cognac over the sugar, and, with all lights out, put a match to the cognac-soaked lumps. When the flames died down and the sugar had melted and dissolved, the glögg was ready. I filled the last glass exactly when midnight struck, and everybody toasted the new year.

In early January Margareta had to return to the teaching position she had held in Stockholm since 1970. I didn't know when I would see her again. Would she want to return? Would I see her in the summer? I didn't dare ask, afraid her reply might not be yes.

During Christmas Margareta had seen Don Hackel, the pathologist who had helped me discover diabetes in the Egyptian sand rats. Hackel knew that Margareta had diabetes and described a summer camp for children with diabetes that physicians at Duke organized in the North Carolina mountains every summer. I thought that perhaps the diabetes camp might entice her to return to Durham. Shortly after she left I called a diabetes specialist who worked at the camp, Dr. Delcher. He expressed interest in Margareta, saying he preferred camp counselors who themselves had diabetes because it was helpful for the diabetic children to see normal-appearing adults who had the same problem.

That spring Delcher invited Margareta to join his camp as a counselor. She wrote to me that, in addition to working at the camp, she would have time to be in Durham with me. I had something wonderful to look forward to.

Matters of Scale

$$W_{HEN}\ M_{ARGARETA}\ returned$$

WHEN MARGARETA returned to Sweden, I resumed my working routine. I tried to find time for writing a textbook in animal physiology but made slow progress. I wasn't satisfied with existing texts, which either dealt with human physiology or, if they were about animals, didn't say what I thought was important for students to understand.

Writing a textbook was far more demanding than I had foreseen. I had good notes from my many years of teaching, but when I started writing I found that my knowledge of physiology wasn't as solid as I had thought. I constantly had to look up information and learn new material, a situation parallel to my early days of teaching, when I had to tackle physiology from a new perspective and experienced a quantum jump in knowledge and insight.

Now, twenty years later, I had a much better grasp of the subject, and my research experience covered a range of fields. However, as I started writing about familiar material, I again had to go back to the original literature to look up what I thought I knew well, to find or draw suitable illustrations, and, more than anything else, constantly verify facts. I now went through a second quantum jump, similar to my experience years before.

I deliberately chose not to look at other textbooks to see what they covered or how their material was organized. I had often found that errors in one book were copied and repeated forever in subsequent books. Besides, I wanted to introduce physiology the way I think it is most interesting, not as others had done it before me. As it turned out, I was just starting a three-year process; the book wasn't published until 1975.

That spring I wrote to Margareta about my work, about bicycling to Duke, about friends I saw and what I had for dinner, about records I played and my work in the garden. I wrote less about how much she meant to me. Our time together at Christmas had again shown me how immensely compatible we were. But Margareta was only twenty-five, and I didn't dare believe she could be seriously interested in a much older man who had two unsuccessful marriages behind him.

Margareta's letters likewise told me about her daily life; at the end of most school days she was exhausted and barely managed to get home on her bicycle. Nevertheless, she went out with friends on weekends. Small things kept us connected.

Margareta arrived four weeks before the diabetes camp in order to attend the extensive training for counselors at Duke Hospital. This time she flew into Washington to avoid the hassle in New York. I met her at the airport, and we drove to Durham. Each morning I woke Margareta with a cup of strong tea with milk. After work we spent as much time together as possible. When Margareta had any time free, she came to see me in the laboratory. She was always easy and relaxed, interested in what we were doing and curious to know more.

Margareta's two weeks at the camp seemed much too long, and when she finally returned we had only a couple of days before she had to return to Stockholm. On the way to Dulles International Airport and while waiting for her departure, we said little. When we walked toward the gate, I sensed that Margareta was near tears. I gave her a hug and watched her walk quickly to the exit door, where she disappeared without turning to wave.

My world felt empty as I drove home to Durham. I sensed that Margareta's feelings were more serious than I had dared believe. It helped to realize that in September I was going to a meeting in Italy, with a stop in Stockholm on the way.

O N M Y W AY to Stockholm all my thoughts were on Margareta. I was uncertain about our relationship. Would I fit into her life as readily now, when she was among old friends?

My plane arrived in the morning, but I had to wait until late afternoon, after school, to see Margareta. We hugged each other and sat talking rapidly, brimming over with happiness, as we drank cup after cup of tea. She wanted me to come to her class the next day; her children already knew about the visiting American professor and were expecting him. I was uncomfortable at the thought of meeting a whole class of nine-year-olds, but I wanted to see the children Margareta had often written about during the past two years.

I sat far back in the classroom, trying to be inconspicuous. The children soon stopped turning around to look; after the first few minutes they were less curious about me than I was about them. I had expected that my presence would distract them, but they went about their work conscientiously. Margareta spoke gently to them, in a kind, warm voice. It was her third year with these children; in grade school in Sweden a teacher usually teaches the same children for three years, from first through third grade. Margareta's girls and boys knew her well and obviously liked her. How different from my teachers in school, I thought; I had either been afraid of them or hated them, or both.

For two days Margareta and I spent much time together, taking long walks in the neighborhood or visiting the school when no children were around. During our long talks I turned over in my mind questions I didn't know how to ask. I wanted to know when we could see each other again. Margareta had friends her own age, and I wondered how important they were to her. Would she spend Christmas with me again?

What would she do the next summer? Would she continue teaching, starting over again the next fall with a new set of first-graders? How could I phrase my words and sound natural? We talked about everything else, but I was too anxious to bring up these questions. Much later I understood that Margareta's uncertainty was as great as mine. How could she know about my feelings for her when I said nothing?

After a short weekend in Stockholm, I took the train to Gothenburg to visit my brother, Patrick. I was terribly upset because I had no clear idea of what Margareta felt or wanted for her future. What could I expect? Would she want to see me again? She was the most lovable woman I had ever met, and I was in turmoil. I called Margareta from a pay phone. Frantically I asked if I could see her before I continued to the meeting in Italy. "Of course," she said softly. "You know you may come any time."

The next day I was back in Stockholm. We sat on a bench outside her house as I tried to tell her why it was so urgent for me to talk with her. I wanted to know if, as soon as I had left, she would prefer the company of others more her age, and I was unable to ask. I was incoherent, but Margareta was calm and lovable as always. Although she didn't say so, I sensed she felt committed to me. I had only a few hours with her, but I left Stockholm feeling reassured.

After my meeting in Italy, I flew to London with Trevor Shaw, a distinguished English neurophysiologist and fellow of the Royal Society, who had also been at the conference. I knew Trevor well because he had served as a visiting professor at Duke in 1966. He showed me his research laboratories at Queen Mary College, then took me home with him to the small village north of London where he lived. Such hospitality is common among academics and a highly meaningful aspect of our lives. In the relaxed setting of a colleague's home one gets to know different and more personal sides of people's lives.

I felt a close friendship with Trevor after spending the evening with him, his wife, and their two daughters. The next morning Trevor drove me to the station for my train to Cambridge. As the train pulled out, I saw him standing on the opposite platform, waiting for the London train due a few minutes later.

I was in the zoology department at the University of Cambridge, talking with Torkel Weis-Fogh and John Treherne, when a telephone call came through with the shocking message that on his way to London Trevor Shaw had fallen from his still-moving train; another train moving in the opposite direction had hit him and instantly killed him. Trevor had been one of Treherne's closest friends.

The Hertfordshire police had found out that I was the last person to see Trevor before the accident and wanted me to attend the inquest. I couldn't stay, so they sent a constable to talk with me in the evening at Torkel's house. He understood that we were terribly upset, and he listened carefully to all I could tell him about my travel with Trevor; I had been with him almost continuously since we left Rome. I told him what was said the last morning, how we got to the station, and how Trevor was waiting for the London train when I left for Cambridge. The constable wrote down an abbreviated statement and asked me to read it. His report was accurate, kind, and considerate.

WHEN I RETURNED to Durham in late September, I devoted myself to work on my textbook. I had finished ten chapters on subjects such as respiration, blood and circulation, energy metabolism, temperature regulation, and water and excretion. The remaining three chapters would cover muscle and locomotion, sense organs, and coordination of function by the nervous and the endocrine systems. These subjects were less familiar to me, and I made slow progress.

Writing is hard work for most scientists, and few write well without putting immense effort into it. I found it difficult merely to start; with academic duties, the laboratory and research students, other manuscripts, correspondence, grant applications, and reports, it was so easy to find excuses to do other, more "urgent" things. The first draft inevitably needed revision and reorganization. Then I revised and rewrote for clarity, put the manuscript aside for some weeks, read it again, and made yet more changes. By the end of the fall term I had made progress on the difficult last three chapters and begun to see the end of the struggle.

Later that year I received in the mail a small book printed in strange characters that had no similarity to anything I had ever seen. Facing the title page, on the left side, it said in ordinary print "Malayalam. Animal

Physiology. Author: Knut Schmidt-Nielsen." This was yet another translation of the small book I had published years before, which had already been translated into a number of languages, including Japanese and Hebrew and Hindi. Malayalam was the twelfth translation, and the publisher hadn't even told me.

I had never heard of Malayalam, so I turned to the great *Merriam Webster's*, my favorite dictionary, where Malayalam was listed as "the Dravidian language of the Malabar coast of India, an offshoot of Tamil dating from about the 9th century." This told me that it is one of the many Indian languages, spoken on the west coast, south of Bombay. I put the strange book in the bookshelf among the other translations, pleased to know my books were read all over the world and not only in English-speaking countries.

Not long afterward I was in Boston to lecture at Harvard and used the opportunity to visit Bent, who was doing graduate study at MIT. He showed me his laboratory and explained his studies of how snake venom affects the acetylcholine receptor of fruitflies. When I was ready to leave, he told me that outside the main building of MIT I could catch a bus that ran straight down Massachusetts Avenue to Harvard Square.

As I came out, a pleasant-looking young man was waiting at the bus stop. He looked as though he was from India, probably a foreign student at MIT. I asked him about the bus, and he replied with a slightly British accent that it was indeed the right bus for Harvard Square. A few minutes later a crowded bus pulled up, and he and I sat down in the last two available seats. After a couple of minutes I started a conversation with some trivial remark about the weather.

"You must be Norwegian" were his first words. I was amazed. Most Americans recognize my accent as foreign, some ask if it is German, and almost nobody identifies it as Norwegian. Many non-Americans do not even detect my non-English background, and I was astonished that a person from India had placed it correctly.

"How in the world do you know?" I asked.

"Oh, my father was Indian ambassador in Oslo, and I lived there for several years." I told him I was a visiting physiologist, and he said he was a graduate student in theoretical physics at MIT. I asked where in India his home was. "Do you know India?" he asked, and I replied that

I had only briefly visited the northern parts of the country, Calcutta, Benares, Delhi, and on west to the deserts of Rajastan. "Well," he said, "I am from much farther south, in fact somewhat south of Bombay."

"Then I assume your native language is Malayalam," I commented casually.

He looked astonished, speechless. Right then we arrived in Harvard Square, and we set off from the bus in different directions. I never knew what the young man thought about a physiologist from Norway who was so unexpectedly familiar with the geographic distribution of the dozens of languages spoken in India. It was an unlikely set of coincidences—his recognition of my nationality and my knowledge of his native language. Since then I have learned from a teacher of fiction writing that some real events are so wildly unlikely that, if used in fiction, they sound too implausible to ring true.

DURING THAT FALL Margareta and I wrote more and longer letters than we had during the entire preceding year. Every day I waited impatiently for the mail to arrive to see if there was a letter from her. I asked her to join me for Christmas, and she agreed.

On December 16 I stood in the crowded arrival hall at the airport and saw Margareta coming out from the customs area. She waved as she pushed her way toward me. I put my arms around her and looked down at her face; she looked up and smiled.

Margareta liked the weather in Durham, sunny and not very cold, unlike the dark winter she had left in Sweden. She had come to believe that winters in North Carolina were always moderate and mild, with no need for warm clothing. We played records, talked, walked in the forest, and went on drives in the countryside. Most weekdays I went to work, and Margareta came with me. She spent long hours in the library or talked with my research students or the postdoctoral associates in my lab. We were sad that we couldn't remain together for more than the few weeks of vacation, but Margareta had to return to her school, and I had my job at Duke. However, this time we knew there was nothing we wanted more than to be together. Margareta visited two schools where she might teach if she were to move to Durham. One was the Friends School, which I had helped support financially when it was founded in

1964. The other was the Duke Preschool, an experimental school affiliated with the psychology department at Duke.

When Margareta left in early January, we had plans to meet in Oslo during her brief winter vacation in February and in Israel at Easter time. Finally, we planned to spend a long summer together, driving around Europe and seeing friends. And after the summer, perhaps Duke Preschool would accept Margareta as a teacher.

During our times apart we continued writing long letters every day. Each of us has kept the other's letters. From January until April 1973, when we were in Israel, I wrote more than sixty letters; in May and June I wrote almost fifty. Between office work, writing my book, and going back and forth between home and the university I did little else that spring. Margareta wrote as much as I did, typing on large sheets of bright yellow paper. Her neighbors thought she was writing a book, for every evening they heard the clatter of her typewriter for hours on end.

Knowing that I soon would see Margareta, I concentrated on my work. For years I had been intrigued by problems related to the size of animals. For example, why are there no insects the size of elephants, or even cats? A flea can jump a hundred times higher than its own height; why can't we? Why are the smallest mammals and birds—shrews and hummingbirds—the same size, about three grams? Is it a coincidence, or is this the smallest size possible for a warm-blooded animal? Would a smaller animal be incapable of producing enough heat to keep warm? How does the size of an elephant's bones differ from the size of a mouse's? All these questions belong to the field of scaling, that is, the significance of size.

Scaling in animals has parallels in engineering, where problems of scaling are so important that they constitute a major subfield. Think about constructing ever-longer bridges or taller buildings. If an engineer were to build a skyscraper from the same materials as a small house, say, bricks, the weight of the upper layers would crush the lower layers because the compressive strength of bricks is insufficient to support the immense weight. Instead, the engineer constructs skyscrapers of reinforced concrete, supported by steel. Similarly, an earthworm doesn't need a rigid skeleton, but an elephant without a bony skeleton would collapse into a flat blob.

Next, consider an animal's need for oxygen. A single-celled amoeba can obtain enough oxygen simply by diffusion, but a mammal needs special respiratory organs and the help of blood and a circulatory system. In a similar way a skyscraper must transport the people who enter its doors by operating fast elevators, for climbing the stairs to the top floors would take hours. Understanding the mechanics of size when it comes to the strength of brick and steel is fairly straightforward; understanding the effects of body size on animal locomotion is more complex, requiring a knowledge of both mechanics and mathematics.

An effective way of making progress with problems that require a multidisciplinary approach is to hold a conference of scientists from related fields, in this case a broad range of mathematicians, engineers, physiologists, and biologists.

Three persons at Duke keenly interested in scaling and locomotion— Vance Tucker, Steven Vogel, and I—decided to arrange a workshop on scaling. We invited Dick Taylor and his Harvard colleague Tom McMahon, an engineer working in applied mathematics. The National Science Foundation declined my application for funds, but with support from the Cocos Foundation of Indianapolis we could invite McNeill Alexander from the University of Leeds, the foremost student of animal mechanics; the eminent muscle physiologist Douglas Wilkie, from the University of London; Torkel Weis-Fogh from Cambridge University, now recovered from his automobile accident; and Sir James Lighthill, a distinguished mathematician from Cambridge, widely recognized for his theoretical work on problems of flying and swimming. These scientists were, on a worldwide scale, eminently suited for the small working conference I had in mind. I especially wanted Lighthill to participate. He would probably decline unless Torkel helped convince him of the importance of our meeting, so I scheduled a trip to Cambridge in February 1973 to see Torkel.

In Cambridge Torkel arranged for me to stay in the fellows' guest room at Christ's College. My bitterly cold room was in a centuries-old building with thick stone walls. The bed had ample thick covers, but if I crept down among the icy sheets, it would take hours for me to get warm. Instead I filled the bathtub with water as hot as I could stand, and after fifteen or twenty minutes in the tub I retreated to the now comfortably cool bed, where I slept all night, warm and snug.

In the evening I dined at high table, where the fellows were engaged in a heated discussion of a request from the students for permission for women to stay overnight in college. A couple of the fellows were unsympathetic, not for reasons of morality, but because they didn't feel that the students' girlfriends should have rights that faculty wives didn't have. "Does a change like this mean that contraceptive information should be provided?" another fellow wondered. "It would seem to be a moral obligation if women are permitted." As if the students don't already know about contraception, I thought. One hard-core oldtimer wanted no women at all in college. "If we begin, we will end up providing midwives as well," he argued. "There are whorehouses and there are colleges, and they are not the same!"

Torkel and I met with Lighthill the next morning. Torkel convinced Lighthill of the importance of our plans, and when we left there was no doubt that he would come for the conference.

I headed for Oslo, satisfied that our conference on scaling would be useful. There I impatiently awaited Margareta's arrival the next day from Stockholm. Our week in Oslo was Margareta's winter vacation, and we enjoyed every minute, going to museums and on long walks in the snowy surroundings. Less than a block from our hotel were the cathedral and the large market square, Stortorvet, where I bought beautiful red tulips that brightened our room. The days went by too fast, and when Margareta's train pulled out from the station, I remained standing on the platform, feeling lonesome and empty once again as it disappeared down the track.

For Easter break that year, Margareta and I arranged to meet in Copenhagen and continue together to Tel Aviv. Margareta bought an inexpensive three-week excursion ticket for me for the Copenhagen–Tel Aviv stretch, an option not available in the United States. For herself she bought a one-week ticket, since that was all the vacation time she had.

On my way to Copenhagen I stopped in New York. I had been asked to serve as a consultant for Biomedical Sciences, a small, innovative company that was developing new products for medical use. A professor at the Albert Einstein College of Medicine had recommended me to the company president, who wanted me to see their research laboratories.

The company swept aside my questions about how to reach the factory in New Jersey; they would pick me up. At the appointed time a long, black limousine pulled up, driven by a very formal-looking chauffeur. As the luxurious vehicle picked up speed, it moved softly over potholes like a huge ship swaying in the ocean swells. It took about an hour and a half to reach the factory. The chauffeur politely declined all my attempts at conversation, and I couldn't extract any information about the company or its president.

At the factory I went through security checks even more impressive than the limousine. For a university professor used to open doors and accessible laboratories, the precautions against industrial espionage seemed extraordinary. Even though I was expected, I was admitted only after careful checks. I signed for a huge identification badge, which I wore for the day and returned against a written receipt when I left.

The young president of the company, Berel Weinstein, greeted me with a broad, informal smile. He was much younger than I expected, dark-haired and slightly built, with intensity in every movement and word. He walked ahead of me with short, quick steps, as if impatient with everything around him. He spent several hours showing me the research laboratories and the factory, talking freely about the products they were developing and his ideas for future projects. He wanted me to see everything and without hesitation answered all my questions in detail.

One of their products was an inexpensive, disposable thermometer, already in production, and many other products were under development. Weinstein was exceedingly knowledgeable, brimming with ideas, and with a flair for problem-solving. He was especially interested in developing quick, disposable methods for blood-urea, cholesterol, and glucose testing, all of which had great commercial potential. He had an excellent understanding of the underlying principles, and there was no end to his plans for additional products. I thought what a great scientist this brilliant young man would make.

I was amazed at how freely he spoke about new products and their underlying principles, a surprising contrast to the strict security measures. He spent hours filling me with stimulating information, and then his company paid me a consultant's fee of $500. I didn't know why they should pay me for the exciting day. The fee was ten times what I received when I worked much harder as a panel member for the National

Science Foundation. Still, I felt that as a federally funded researcher, I shouldn't accept pay from a commercial company, and when I returned home I turned the check over to my department at Duke.

In the late afternoon the elegant limousine returned me to Manhattan. I leaned back among the comfortable cushions and noticed a rose in a small vase. I checked my spacious surroundings and found candies, cigars, and a well-stocked bar. Life in higher business circles was a new experience, the advantages of which came into sharp focus later in the evening as I sat cramped in the tourist-class seat in the plane to Copenhagen.

Our flight to Tel Aviv was via Amsterdam, where security measures were extraordinary as a result of several attempts at hijacking and sabotage. All checked baggage for our flight was assembled in a room where Israeli security agents asked all passengers to identify their luggage. We pointed to ours and watched while every item in every suitcase was removed, scrutinized, felt and pressed and pinched.

While we waited for our flight, we looked out over the dark airport. In the floodlights we saw that we were guarded by two light tanks armed with anti-aircraft artillery, as if the authorities were ready for war. Our plane wasn't permitted to taxi to the departure gate; we were taken by bus to the plane, which was parked on the other side of the airport. When the bus started for the plane, it was accompanied by three armored cars with soldiers who carried heavy automatic weapons.

We finally departed at four in the morning. Margareta and I sat close together under a blanket, much too excited to sleep, happy just being together. At the airport in Tel Aviv, we were welcomed by Amiram Shkolnik, who had been my guide since my first visit to Israel many years before. As we left the airport, we passed orange groves, breathing in mild air filled with the heavy fragrance of orange blossoms. Then we drove north to the Kibbutz Kabri in the western Galilee, and after two almost sleepless nights on planes I slept most of the day with Margareta next to me.

During the Easter week, Amiram showed us many places we knew from the Bible. For Margareta this was especially exciting because she was in the middle of teaching her schoolchildren about Palestine at the time of Jesus. We went to the Lake of Galilee, where Jesus fed five thousand people with five loaves of bread and two fish; and to Cana, where

Jesus turned water into wine. At Akko, the immense fortifications built by crusaders reminded me of my early history lessons. The Norwegian king Sigurd Jorsalafar, a renowned crusader, had come to the Holy Land in 1110 with a fleet of fifty-five ships to help free the Holy Land from the infidels. Jorsala was the Old Norse name for Jerusalem. Were we looking at fortifications build by a Norwegian king, I wondered.

In Jerusalem we walked all over the old city. The air was thick with incense and exotic spices, and the narrow, winding streets were crowded with Arabs, Jews, and tourists. We drove to Masada, the ancient and unapproachable mountain fortress that is a national shrine for modern Israel. The nearly flat mountaintop, several hundred meters above the Dead Sea, drops off vertically on all sides. Around the year 70, a group of Jewish zealots held out for two years against a Roman army of 15,000; the story goes that the last Jewish defenders committed suicide, preferring death to enslavement.

At the end of the week Margareta returned to her school while I remained in Israel as a visiting professor at the University of Tel Aviv. I lectured and conducted student seminars almost every day.

Ten or twelve students in the zoology department were studying the physiology of desert rats, lizards, turtles, and other local animals. During my seminars one of them would present his or her research plan and describe the progress made. They were nervous about performing in front of an American professor and came well prepared with slides and manuscripts, but when I involved the other students in a discussion of what we had heard, they forgot about manuscripts and slides and were soon engaged in a lively discussion. I could stay in the background and let the students carry on.

After my last lecture, the research students arranged a small good-bye party for me in the laboratory to express their appreciation for our time together. They told me that my visit had been useful and stimulating, not realizing how actively they themselves had contributed. I, in turn, was grateful for having eager and interested students who wanted to listen and learn, more so than any other group of students I have lectured to.

Afterward Amiram drove me to the airport, where I again encountered extraordinary security measures. The flight was uneventful until we were approaching Copenhagen, when the plane suddenly started violently pitching and yawing. Obviously, the plane had been sabotaged and would hit the ground in a violent crash. My thoughts were on Margareta; we would never see each other again. Then, equally suddenly, the plane was back in a steady forward flight. My palms were wet and my heart was pounding, but I could breathe. Two young men across the aisle from me were chalk white. "That was close," I said, trying to sound casual. They nodded but said nothing. No explanation was given for the unusual movements. I asked one of the cabin crew, but she had evidently been instructed not to discuss anything that might have a bearing on safety. Ten minutes later we were on the ground, and I boarded the plane for Stockholm.

I brought with me a gift for Margareta; it was a ring with a large, clear stone mounted in a striking gold setting. She was amazed that I knew her exact ring size, although that had been easy to find out. Before she left Israel I had asked if I might look more closely at a ring she was wearing. She handed it to me, and it fitted perfectly on the little finger of my left hand, giving me the precise measure I needed. Before I returned home to Durham we spent a long and happy weekend, talking about our plans for the summer.

In the late spring Margareta received an offer of a position at Duke Preschool, and in the fall of 1973 she came to Durham to teach children of about the same age as those she had taught in Stockholm. We rarely went out for dinner; it was easier to arrange meals at home suitable for Margareta's diet, and we preferred the quiet in each other's company. Margareta, who is an excellent cook, eventually took over the kitchen. She liked to experiment, and the result was a considerable increase in the amount of fresh vegetables we ate. At dinner she enthusiastically told me about the day's work and the fun the children had while learning. I was amazed at how different this school was from what I remembered from my childhood.

THE CONFERENCE on problems of scaling was convened in late September. Each morning one of the nine participants gave a lecture that was open to the university community, and after lunch we had another lecture. A keenly interested group of forty to sixty students and faculty members attended regularly. Our intensive working sessions were sandwiched between the lectures.

I had invited Professor Lighthill to give a formal evening lecture, which a couple of hundred people might attend. He agreed to talk about the fluid dynamics of swimming. Lighthill is an intellectual of extraordinary caliber, elected to the Royal Society at the age of twenty-nine; he is also physically imposing, with a huge body and an impressively large, round head that towers over most others.

From his position on the podium Lighthill discussed hydrodynamics in clear and simple terms and described various types of swimming movements in fish. Then suddenly this large man ran across the stage with small, feather-light steps, one hand wiggling back and forth behind his bottom to show what is meant by carangiform swimming, which is how an ordinary fish swims. The audi-

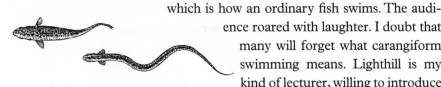

ence roared with laughter. I doubt that many will forget what carangiform swimming means. Lighthill is my kind of lecturer, willing to introduce humor to make a point. However, he did not demonstrate anguilliform swimming, which is how eels swim, perhaps because his bulk precluded a good imitation.

Having had to persuade Lighthill to come to our conference, I was apprehensive about his reaction. It was therefore marvelous to hear him casually remark on the last day that he now wanted to organize a major international conference on scaling in animal locomotion.

Dick Taylor wrote that for him the conference was especially useful in deciding the direction of his future research. "I came back feeling that I can still make a major contribution toward understanding locomotion and seeing clearly the kinds of questions that really must be answered," he said. His later work on the importance of body size in animal locomotion has indeed been distinguished.

That fall Margareta signed up for a sculpture course. She suggested that it would be more fun if I also took the course. Sculpture was the last thing I thought I could do, but I called the teacher, Martha Wittels, and asked if someone without any talent might take her course. "Oh, that is precisely the kind of people I want," she answered. "I love to have people I can really do something meaningful for."

An early assignment was to learn about the human head by first copying in clay a human skull and then covering the skull with skin, filling in the eyes, and putting on hair, all in clay. Helped by my knowledge of anatomy, I had no difficulty with the skull. The finished work was a long-nosed man with full lips who might have been a middle-aged Spanish nobleman.

My next work was a large frog with bulging eyes. Martha didn't like it, but I am happy I kept the piece. It now sits on our mantelpiece, where it carries on a friendly conversation with a huge wooden frog from Japan that a Swedish friend, Stephen Thesleff, gave me when he worked in my laboratory, studying the muscle physiology of crab-eating frogs.

After some months I started working with wood while Margareta switched to stone. My first piece was so bad that it went right into the fireplace. Another failure sits in a corner, where it neither satisfies nor bothers me. However, some people have said they like it, even without knowing who made it.

I wanted to work with maple for my next piece, but after starting I found that the wood my colleague Peter Klopfer had given me from his woodpile was oak. I soon discovered that oak is good to work with but cracks badly. I used chisels to remove chips until I had what I intended, a man and a woman standing close together, arms around each other. I left the wood untreated, and it was in fact a better choice than the maple I had wanted. Martha liked my sculpture and was pleased; with a smile she suggested that the woman perhaps was red-haired. I was satisfied that I could produce an acceptable piece of sculpture. However, I have not continued sculpting; there are too many other things I want to do.

Our small conference on scaling proved to be a more important catalyst than I had anticipated. The two years after the conference saw significant developments and the emergence of a new research field. Lighthill's conference in 1975 was attended by scientists from around the world, who lectured on problems of terrestrial, aquatic, and aerial locomotion. They presented new data on muscle function and skeletal systems, mechanics and fluid dynamics, oxygen consumption and energy supply, always with a view to animal size and possible critical boundary conditions. All the participants left with new ideas and new friends, scientists from around the world we knew only from their publications. These new personal contacts remained important for all of us. The proceedings of the conference were edited by Tim Pedley and published under the title *Scale Effects in Animal Locomotion.* This volume became a milestone in a rapidly developing field.

The interest in scaling has continued to increase. I have lectured on scaling at universities around the world and have given courses in the subject. In 1984 I published a book, *Scaling: Why Is Animal Size So Important?* At about the same time, my former student Bill Calder published the book *Size, Function, and Life History.* Although Calder's book deals with many of the same problems that I discussed, his provides better coverage of problems of ecology and natural history; in many ways the two complement each other.

Shortly after the Cambridge conference, Torkel Weis-Fogh wrote me about his plans for a sabbatical leave for the coming calendar year. Before spending a couple of months at the California Institute of Technology, he wanted to spend a month with me at Duke. After California he would go to Hawaii and study the swimming of fish. Finally, before returning home, he would meet with several groups of Soviet scientists who were working on flying and swimming. Torkel also described his plans for a book called *Animal Flight: A Study in Ultimate Function,* which would include structural topics, muscle physiology, fuel supply, special metabolic systems, and perhaps problems in sensory and control systems. "All these functions have been brought to a peak in flying animals," he wrote, "hence the use of the word 'ultimate' in the proposed title."

Neither travel nor book materialized. Torkel committed suicide on November 13, 1975. His close friend Dorothy, whom I had met at his house, called me from London to tell me of his death.

Later, John Treherne gave me the details. Torkel's friends and colleagues knew that he continued to suffer from the injuries to his rib cage and left lung, and that he also was overworked and emotionally unstable. When he failed to appear at his office, his house was searched. He was found dead in a secluded place, as if he had intended to hide. How characteristic of Torkel's personality, Treherne said, that he concealed himself in death as he had also concealed his feelings in life, always presenting a rational surface to the world.

Torkel's suicide was not the first among my friends; the closest was my former chairman and good friend at Duke, Ed Horn, who in 1969 had killed himself with cyanide. Ed had been of immense help to me during the difficult years after my divorce in 1962.

I had last seen Torkel at the scaling conference in Cambridge. After its official close I spent the night at his house, and he spoke about having once contemplated suicide but assured me that these thoughts were in the past. He sounded well balanced, warm, and human as he talked openly about himself. We were later joined by his friend Dorothy, a distinguished social scientist at Imperial College, London, and had a pleasant dinner and evening.

Torkel's death was a tragic loss to biological science. His contributions to our understanding of insect flight ranged from the flight of the migratory locust to the hovering of an insect in still air. He showed that insects smaller than about two millimeters cannot use the same methods for hovering as larger animals because the viscosity of the air has an effect that, for tiny animals, can be likened to that for a human to swim in molasses. Torkel discovered principles previously unknown to fluid dynamicists.

After Torkel's death, I wrote to Dorothy to tell her what Torkel and I had talked about the last time I saw him and the grief we all felt. Dorothy sent me a long reply. "Torkel did not have many friends as I know friends," she wrote, "and he always spoke of you personally and scientifically in a way that made me wish to know you, even before we actually met." She felt that Torkel had suffered from a true clinical depression. He had been most depressed the previous Christmas, she said, and afterward his doctor and drugs helped him through the teaching term.

Now I understood his sudden change of plans before Christmas a year earlier, when he had planned to spend Christmas in Durham with Margareta and me, celebrating a real Scandinavian holiday. Shortly before his expected arrival, he had telephoned to say that he was unable to come, but without explaining why.

I was grateful when Dorothy again wrote to me, and in a long reply I wrote: "A suicide is always difficult to accept, not only the bereavement, but the inevitable feelings of guilt. During the last several years five or six friends, colleagues, and a former student have taken their own lives. Is the feeling of guilt inescapable? When Torkel said that his thoughts of suicide were over, I believed him, and I feel guilty for believing him."

In her next letter Dorothy wrote: "As you say, we barely know each other, but I am only trusting the strong feeling I had *whenever* Torkel spoke of you, that you were—and I think I can say it—almost the only *close* friend he had.

"But then Torkel was one of the most lonely persons I have met. You may shed some light on this.—And we will all feel that it was such a waste—as James Lighthill said, 'Scientifically things will not now be done that would have been, simply because there is no one else to do them.'"

I had thought a great deal about Torkel, both before and after his death. To me it seems that he went wrong in basing his life entirely on rational analysis of all situations. Because Hanne had had the ability to fit into his life exactly, he never discovered that his emotional needs had been met. Since the accident it seemed that he had tried to base his life entirely on making only rational and logical decisions about very important personal relationships.

Part IV

Measuring
a Life

[1975–1995]

North of the Arctic Circle

EARLY ONE October morning in 1975, as I was bicycling to work, a freak mechanical failure blocked the front wheel. The bicycle stopped instantly, throwing me forward. I reflexively grasped the handlebars and was hurled to the concrete pavement face first.

I remember the very moment I hit the ground and the sound of crushing bones. I never lost consciousness and gratefully concluded I had no severe concussion. I knew that I had lost several teeth, that my nose was broken, that the bones of the maxillary sinus on the right side were smashed, and that my mandible was broken—I could hear the grating sound of crushed bone when I tried to move my lower jaw.

I was sitting on the ground, hoping for help. It was early, about seven, and the only car that came by didn't stop. Shortly afterward a police car drove up; probably the car that passed me had reported the accident to the nearby campus safety office. The policeman wanted to call an ambulance, but I convinced him that I would get to the hospital faster if he would just drive me there.

After what seemed like hours in the emergency room, a sympathetic resident or senior medical student rolled me off in a wheelchair to have X rays taken. He looked relieved when I mentioned the name of a surgeon, a former neighbor, who I thought would be helpful.

Much of the day I spent sitting alone in the emergency room in my wheelchair, while at intervals other physicians came to examine me. Each time I was asked to open my mouth, the pain was excruciating, and I hated the sound of broken bones as I moved my jaw. I didn't understand why so many needed to examine me, and when yet another asked me to open my mouth, I asked, "Is it necessary?" He shouted some angry words and stalked out. At least no one bothered me until I was moved to an operating room toward six in the evening, more than ten hours after the accident.

I woke up with my jaws wired together. A nurse explained that this would hold the fractured lower jaw immobilized until it healed. She gave me a pair of wire-cutters as an emergency measure in case I vomited. The wire-cutters hung in a band around my neck for the next two months.

The next day there were more X rays and exams, but at least no one asked me to open my mouth. These X rays revealed an additional fracture of the maxilla, but that seemed irrelevant because my jaws were already firmly immobilized.

Nobody discovered that my right shoulder was also fractured. My face received all the attention, and other bruises didn't seem important. It wasn't until a couple of years later, when my shoulder became increasingly stiff and painful, that I had it examined and the healed fractures showed up on new X rays.

On the second day in the hospital the surgeon who had wired my jaws reluctantly discharged me. Margareta had convinced him that she could take care of me better than the people in the hospital could. I was immensely grateful for Margareta's affectionate caring, but I suffered some agonizing days while the worst pain slowly abated.

For two months I consumed only liquid food, which I drank through a straw. Now I made an interesting observation. Once food was homogenized, whether it was chicken or beef or anything else, it all tasted the same. I realized that texture is far more important to what we call "taste" than I had thought. The Japanese understand this; they consider the texture an extremely important component of food, an aspect that Westerners rarely pay any attention to. I was never hungry, but I knew I needed food and forced myself to drink. I was eagerly looking forward to the moment the wiring would be removed and I could again eat like a normal human being.

When the wiring was removed, the surgeon warned me that I might have trouble opening my mouth fully, but I had not expected to be unable to open it at all; it was as if my jaws were glued together. I couldn't put even the smallest bite of food into my mouth. I practiced using a pencil as a handy measure to gauge my progress. After days and days of painful work I could open my jaws just enough to slip the pencil in between my teeth. After weeks more I could eat small pieces of food. A year and a lot of dental work later, I considered myself fully recovered.

In spite of the long recovery period and all the pain, I knew I was lucky that I had not been killed. I wrote to an engineer friend that if a person intends to hit his head hard against a concrete pavement, it is safer to use a collapsible and relatively nonvital part of the head, rather than the braincase, which might shatter like an eggshell.

In June 1976, Margareta and I visited family in Stockholm. We had decided to be married, although I had long hesitated mentioning the possibility because of the difference in our ages and my unfortunate record. The turning point came when Margareta one day remarked, "If I ever wanted to get married, it would be to someone like you." This was when it became clear to both of us that there was nothing we wanted more than to be married to each other.

We wanted to be married in Sweden and talked to a minister, not because we wanted a religious ceremony, but because in Sweden the church maintains all official marriage records. After our meeting with the minister, we knew that we wanted a simple civil ceremony. We would be married in Stockholm the following summer.

From Sweden we went to England to attend the centenary celebrations of the Physiological Society. There Jean Scherrer from Paris introduced himself and asked if I would give a plenary lecture at the International Physiological Congress in Paris the next year. The organizers wanted me to speak on the role of physiological science in the future development of the world's deserts.

I explained that if the subject had to be the exploitation of desert areas for the benefit of mankind, I was not the right person to ask. Making the desert bloom was a romantic dream, voiced by David Ben-Gurion when he was prime minister of Israel, but it could not become a reality with the aid of science. On the contrary, attempts at utilizing the world's deserts for the benefit of humans can only destroy what natural deserts are left. I didn't expect to hear again from the congress organizers. However, soon after I returned home I received an invitation that gave me full freedom as to the content of my lecture.

The 1970s were a productive decade. My graduate students and postdoctoral associates contributed greatly to a number of successful research projects. For several years we concentrated on how much energy it takes for birds to fly and on the role of the respiratory system and the blood in supplying enough oxygen. We especially wanted to know why birds can tolerate much higher altitudes than mammals can. How is it that birds, which are heavily dependent on oxygen for flight, can fly about at altitudes where humans and other mammals can scarcely move for lack of oxygen?

I was also interested in how birds eliminate the excess heat they generate during flight. Our wind-tunnel studies showed that gulls experience a tenfold increase in heat production when flying. Since birds have no sweat glands that help them avoid becoming overheated, how do they manage? We found that much of the extra heat during flight is lost from the trailing feet.

In other circumstances, however, birds must conserve body heat. When a duck stands on ice or swims in ice-cold water, the blood that flows to the feet is cooled to near freezing. If the ice-cold blood from the bird's feet were to flow back to the body core without being rewarmed, it would rapidly chill the entire bird.

We found, however, that no such cooling takes place. Birds have an ingenious arrangement of blood vessels that allows the blood to be reheated before it returns to the body. The arteries that supply blood to the feet run along and very close to the veins that return blood from the feet. As the two blood streams flow past each other, the warm arterial blood serves to heat the chilled blood in the nearby veins. In this manner the venous blood is rewarmed before it reaches the body core. Because the arterial and venous blood flow in opposite directions, the arrangement is called a countercurrent flow.

The effectiveness of the countercurrent heat exchange depends on how close the blood vessels lie to each other, their length, the thickness of the walls, and how fast the blood flows. Optimally, the venous blood is warmed virtually to the arterial temperature before it reaches the body and thus has no chilling effect on the body core.

But what happens if a bird is exposed to below-freezing temperatures? It must keep its feet warm enough to avoid frost damage. My associate Del Kilgore found that birds then produce sufficient extra heat to supply the feet with blood warm enough to keep them from being frostbitten. The metabolism is then increased enough to produce the extra heat that is needed, bypassing the countercurrent exchange mechanism.

Two of my students, Mike Fedak and Berry Pinshow, were especially interested in how emperor penguins survive under the harsh polar conditions in which they breed. Whereas most birds breed during spring and summer, the most favorable time of the year, the emperor penguin breeds and incubates its egg during the dark and frigid Antarctic winter. When winter approaches, these birds leave the open water and walk for 50 or 100 kilometers over the sea ice to their rookeries on the permanent ice shelf, where they court and mate. The female produces a single egg, which she places upon the feet of the male, who covers it with a flap of skin to keep it warm during incubation.

The female then leaves and returns to the sea to feed while the male remains, standing on the ice with the egg on his feet, while the air temperature falls far below freezing. He incubates the egg for over sixty days, one of the longest incubation periods known; furthermore, he cannot feed from the moment he leaves the ocean.

Few males in the animal kingdom invest as much in reproduction as the male emperor penguin; many other males contribute nothing more than the insemination of the female. In addition to doing all the incubation and starving while doing so, when the chick hatches the male emperor penguin remains until the female returns to feed the chick and relieve him. If she is late, the male feeds the chick with a milklike secretion from his esophagus. When the female arrives, he finally returns to the sea to feed. Thereafter the male and female shuttle back and forth to feed the chick until it is large enough to undertake the long walk to the sea.

From the time the birds set out for the breeding grounds until the female returns, the male may starve for as long as 100 days. Like other animals, the emperor penguin stores energy in the form of fat; a large male weighing thirty-five kilograms (seventy-seven pounds) when he leaves the sea may have as much as fifteen kilograms of fat stored in his body.

Fedak and Pinshow showed that the walk to and from the breeding ground requires about 1.5 kilograms of fat. While standing alone on the ice in howling winds and temperatures as low as −40°C (−40°F), a bird will metabolize about 0.2 kilogram of fat per day to keep warm. Thus, during 100 days of starvation a male penguin needs at least 20 kilograms of fat, more than is stored by even the largest birds.

Why doesn't the bird starve to death long before the incubation is over? The answer relates to heat conservation: the animals survive because they breed in colonies of several thousand birds huddling close together. Instead of being exposed to freezing air on all sides, as in laboratory experiments, a bird huddled close to neighboring birds loses much less heat.

It was the young French physiologist Yvon LeMaho who measured the effect of huddling. Although we didn't know about it until later, he was in the Antarctic at the same time as Fedak and Pinshow, and he

found that penguins that huddle closely together metabolize only about 0.1 kilogram of fat per day, half as much as they would standing alone. Although the male is quite lean after incubation, he still has enough fat left for the walk back to the sea.

Why the emperor penguin chooses to reproduce during the most hostile season is another question. Most likely the timing allows the chicks to grow and mature during the spring so they are ready to walk to the sea and fend for themselves at the beginning of summer, the most favorable time of the year.

A great deal of work goes into studies like these; at times we must develop new methods and new equipment, and often false leads take us up a blind alley. Finding out, uncovering the problems and determining the answers, is what makes it so exciting to study how animals manage in environments that seem impossibly hostile. And once we know how things work, we discover that the fundamental scientific concepts usually are simple and easily understood.

ALTHOUGH MY interests had broadened considerably since I first set foot in the deserts of Africa and Australia, camels and their ability to minimize water loss from the respiratory tract continued to intrigue me. At a meeting in Switzerland I discussed with Bob Schroter, an engineer at the Imperial College in London, the extremely low evaporation from the respiratory tract of camels, although I couldn't explain the physical mechanism that made it possible. Bob was keenly interested in the mystery, and we agreed to study the problem, arranging this time to work in Kenya.

In May 1977 I flew to Kenya with Margareta. Landing at the Nairobi airport late in the evening, we stepped off the plane and into the velvety tropical darkness, which was permeated by the fragrance of exotic flowers. On our way to the terminal, an official separated us from the other passengers and politely ordered us to follow him. We didn't know why until we entered the VIP lounge and found a group of friends waiting to welcome us: Dick and Ann Taylor and several of Dick's collaborators; Bob Schroter, who had just arrived; Geoffrey Maloiy, the professor of physiology at the Veterinary College in Nairobi, and his wife, Josephine. Dick later told me that Geoffrey had arranged the reception because he considered scientists no less important than all the useless politicians who always receive a VIP welcome.

Geoffrey, a well-known scientist, is also a respected elder of the Maasai tribe. The Maasai are a tall and handsome people who raise cattle and have their own distinct, highly independent culture. At six feet four inches (193 centimeters), Geoffrey towers over most people. I was happy to see him again and to meet the slim and beautiful Josephine. Both she and Geoffrey are fluent in English as well as Maasai and Swahili.

Several weeks later we would encounter Geoffrey and Josephine again at a meeting in Denmark. Geoffrey arrived first, and Josephine was supposed to come a day or two later. To inquire about her arrival, Geoffrey with unaffected grace walked to the reception desk and with the most natural expression politely asked in impeccable English, "Has the African princess arrived yet?" He was informed that, according to a telegram, she would arrive the next day.

In Nairobi Geoffrey had arranged to have three camels from the desert areas to the north waiting for us at the research station at Muguga. Dick, who was running a large research project there, solved my transportation problems. He lived in an apartment adjoining ours, and each morning I rode with him in his jeep to Muguga, along with Bob Schroter, who lived in the nearby United Kenya Club.

The thirty-kilometer road to Muguga runs through undulating land that rises toward the Ngong Hills in the west, just at the edge of the spectacular escarpment that drops off into the Rift Valley. In many places the surrounding land is covered by deep layers of highly fertile, red volcanic soil, which supports local agriculture. Nairobi is located exactly on the Equator, but because of its altitude, about 5,400 feet (1,600 meters), the climate remains moderate and comfortable all year.

All three camels were males. I would have preferred females, which are easier to handle, but here as in the Sahara, nomadic herdsmen are unwilling to sell females because of their value for breeding. One of the camels was enormous, and although our animals were not especially difficult, the animal handlers at Muguga were afraid of them. Fortunately, I had not forgotten my camel-handling skills and could show the men how to lead the camels and restrain them for our work.

The first step was to build an airtight plastic face mask for each camel so that the animal's exhaled air could be conveyed directly to our instruments. We knew from my Australian study in 1962 that much less water was lost in the exhaled air than could be explained. At that time it was generally believed that exhaled air always is at body temperature and 100 percent saturated with water vapor.

To account for camels' lower-than-expected water loss in the exhaled air, we tested two hypotheses: that the air could be exhaled at a temperature lower than body temperature, and that the air could be less than fully saturated with water vapor.

With the masks on the camels, we measured the temperature of the exhaled air with fine wire thermocouples that responded to temperature changes within a fraction of a second. We found that our first guess was correct: the exhaled air was usually much below the camel's body temperature. When the camel inhales, the surfaces of the nasal passageways are cooled by the flow of air streaming over them. When the camel exhales, the air from the lungs is cooled as it flows over these cool surfaces, leaving the nares at well below body temperature. Because the air is cooled, it holds less water than it did when leaving the lungs.

Our results failed to confirm our second hypothesis. One reason may have been that the camels were not fully adapted to desert conditions, since the climate in the Nairobi highlands is not as dry as that in the hot deserts of central Australia and North Africa. Another reason may have been that the hypothetical mechanism we were looking for comes into play only when camels are dehydrated. We did not water our camels, but the animal handlers who walked them from the stables to our working area permitted them to graze on the lush grass along the road. As a result, the animals had far more water than they would have had from the dry feed available in a real desert.

Lacking a clear answer, we decided to measure the exhaled air of other large animals. We examined cows, sheep, goats, wildebeest, and waterbuck and found that all of them exhaled air at about 10°C below body temperature, similar to the camels.

One of the faculty at the Veterinary College, Vaughan Langman, was studying the water needs of giraffes because in nature they seem to have a lower water requirement than one would predict. Langman thought that measurements on giraffes, similar to ours on camels, might help explain their low water needs.

Langman worked with giraffes at a field station near the Athi River, not far from Nairobi. The giraffes needed fresh feed in addition to the grain he usually fed them, and one day Margareta  and I helped cut branches from the surrounding thorny acacia shrubs, the food the giraffes prefer in nature. This took some care, for the acacia has straight, sharp thorns the size of darning needles. I watched in amazement as the giraffes gracefully swung their long, slender tongues around the branches and gently pulled them into their mouths, seemingly without noticing the thorns. It wasn't clear whether the mouths of these animals are insensitive to the thorns, or whether they somehow manipulate the branches within the mouth in such a way that the thorns are never impaled in the soft tissues.

With endless patience Langman had succeeded in taming these nervous animals to the point that they were almost cooperative. He gingerly induced them to enter a narrow chute where he could reach their heads by standing on a platform about four meters above the ground. We climbed up to the platform and came close enough to touch the head of one of the giraffes. Vaughan worked slowly and gently until it was willing to accept a facial mask so that we could make our measurements. The temperature of the exhaled air was indeed far below the body temperature, exactly as in the other large animals we had measured. The results had important implications for Langman, who had been concerned about how young giraffe calves get enough water when, for long periods, they apparently never come near any open source of drinking water.

After four weeks in Kenya, we had secured answers to our first question, the temperature of the exhaled air, and I realized that we were unlikely to obtain useful answers to the second question. We had enjoyed living in Nairobi, having our own household and buying food in the local market, and seeing the magnificent herds of wildlife at several game parks.

IN JUNE Margareta and I flew to Stockholm and visited Margareta's parents in Mörby, one of Stockholm's older suburbs. Their large two-story wooden house, painted the red that characterizes almost every farmhouse in Sweden and Norway, sat on a rocky lot among birch trees and pines. Nevertheless, Margareta's father had an abundance of flowers alongside the house and a bed of roses that bloomed throughout the summer.

After a few more days with Margareta's parents at their country house on the coast, Margareta and I went to Stockholm's City Hall with the required documents to make an appointment for our wedding. Next we went to a jeweler and ordered a ring for Margareta, which we had engraved with our initials and the date, "MC 21/6 1977 KSN."

June 21 is Midsummer Day, the longest day of the year. For Swedes, Midsummer is a major holiday, but it is always celebrated on a Friday, just as in the United States Presidents' Day is celebrated on a Monday, regardless of the actual date. Celebrating Midsummer on a Friday gives Swedes a long weekend to recover from any excesses during the exuberant festivities.

Our wedding was on the real Midsummer Day. We woke to sun and cloudless skies. On the way to City Hall, Margareta, who had forgotten about lunch, suddenly realized that her blood sugar was falling precipitously. We stopped at a small Greek restaurant, but afterward neither of us had any idea of how the food tasted.

We arrived late at City Hall, where Margareta's oldest friend, Ulrika, was waiting for us, slightly nervous. She was the official witness at the ceremony. The alderman, a relaxed young woman, conducted a brief and simple ceremony, and in a few minutes I was married to the most wonderful woman I have ever known. We took Ulrika with us to Mörby, where Margareta's mother had prepared an elegant dinner. It was a wonderfully happy celebration for all of us, far more meaningful than a big wedding.

That evening Margareta and I set out for Norway to visit friends in Tromsö, far north of the Arctic Circle. We had a few weeks to spare before I was to give a major lecture at the International Congress of Physiology in Paris, and what could be better than visiting the coast of Norway?

We took the train from Stockholm and slept well despite the light-filled night. In the morning we were traveling through open, uninhabited country. The train slowed to a stop beside a large sign announcing that we were at the Arctic Circle, 7,389 kilometers, or about 4,600 miles, from the Equator, where we had been only a couple of weeks before, watching elephants and giraffes and zebras. The sign also stated that we were 2,611 kilometers, or 1,600 miles, from the North Pole, although the train wouldn't run that far. A long row of white stones formed a straight line stretching westward as far as we could see, marking the

Arctic Circle. We laughed, wondering whether the stones, after continuing into Norway and to the coast, dipped into the Atlantic and continued along the bottom of the ocean to reemerge on the North American continent north of Hudson Bay.

The train continued through the treeless tundra of Swedish Lapland. We saw a few glimpses of sun between snow flurries and passed a lake that was still ice-covered. Later in the day we crossed into Norway and arrived at the port city of Narvik, the end of the rail line. Narvik's harbor remains ice-free in winter and is the main center for the export of iron ore from the rich mines in Sweden. I had not been there since the summer of 1939, when I was studying marine birds at the island of Röst and encountered the two German spies.

We spent the night in Narvik and left by bus for Tromsö the next morning. On the way it started snowing, but the driver handled the bus skillfully on the narrow, winding road. Six hours later we crossed a long, modern bridge to the island where the city of Tromsö is located. The last time I had been there, the city had had no direct connection to the mainland; all traffic had gone by sea.

Johan Steen, whom we had last seen in Oslo in 1967, was waiting for us at the bus station. He had moved to the newly established University of Tromsö, the northernmost university in the world. He immediately took us to his field station to show us his work on ptarmigan and reindeer. Seeing the reindeer close up, we were amazed at how small they are. We probably have such wrong ideas about their size because they are invariably photographed alongside Lapps, who tend to be quite short.

We heard on the radio that farther north a severe blizzard had deposited masses of snow and large areas had no electricity. On the second day people were still without electricity. "Oh, they are used to it," people in Tromsö said when I asked how people could manage. Midsummer, I thought, and cut off from the world by a blizzard!

In spite of the wind and rain with temperatures barely above freezing, Johan took us to Karlsöy, a remote island where the University of Tromsö has a research station. There several research students were investigating the breeding biology of the ptarmigan and greylag goose under natural conditions. One of the students had determined that the chick of the ptarmigan, while still in the egg, learns to recognize the call of its mother, indicating that those small embryos are more aware than we tend to assume.

Upon returning from Karlsöy we had another day with Johan and his wife, who served us a magnificent reindeer stew. Some ten years later, after the Chernobyl catastrophe, thousands of reindeer in northern Norway and Sweden were contaminated by radioactive fallout and had to be killed and buried.

We left Tromsö, impressed by the new university and what it had achieved, far north of the Arctic Circle.

W̶E̶ STAYED IN Stockholm, then began a roundabout journey to the physiological congress in Paris. After two pleasant days in Denmark we made our way to Saulieu, a small town some 300 kilometers southeast of Paris, for a visit to the ethologist Hilde Gauthier-Pilters. I had first met Hilde in 1954 in Béni Abbès, where she courageously went alone to distant nomad camps and stayed for weeks while observing the behavior of free-grazing camels, making notes about the plants they browsed on, and at the wells measuring the huge amounts of water they drank.

Hilde later married the manager of the research station in Béni Abbès, the botanist J. Gauthier, and settled in France. When Margareta and I visited her, the Canadian zoologist Anne Dagg was there. The two women had collaborated on camel studies in Mauritania and were writing a book on camels, and they asked me to read their manuscript.

From Saulieu we went by train to Goderville, in Normandy, to visit Yvon LeMaho, who had studied the metabolism of emperor penguins during their breeding season. He had spent several months with me at Duke and had become a good friend. Knowing that we would be in France, he invited us to visit him at his parents' home. His father, a rotund and congenial countryman, ran a wholesale wine business on the outskirts of the small town, where the family lived in a large stone house with a sizable garden. Yvon's mother served some magnificent meals, and his father opened some remarkable wines from his own cellar. On our last day he served a bottle of burgundy from 1945, known as the greatest wine year of the century. It was said that in 1945, after five years of German occupation, the grape vines celebrated the liberation by producing a vintage of a quality never seen before or since. The rich and powerful Chambertin from Piat Père & Fils was probably the greatest wine I have ever tasted.

From Goderville we set off for the International Congress of Physiology in Paris, where I had been invited to discuss how an understanding of physiology can aid in the development of the world's deserts and help civilization blossom in arid conditions—a view I neither subscribed to nor wanted to encourage. I was also apprehensive because the president of the congress, Maurice Fontaine, had written that my talk would be one of only three plenary lectures, attended by students, journalists, and interested laymen, as well as all the congress' participants. Fontaine had accepted my reservations about the topic, reflected in the title, "Physiological and Cultural Adaptations to Life in a Desert Environment: Possibilities and Limitations." In my talk I concentrated on limitations rather than on rose-tinted optimism about development.

Deserts and arid lands are indeed important. More than 30 percent of the land area of the Earth is arid or semi-arid, which simply means that nearly one third of all the land has inadequate rainfall. A common assumption is that all problems can be solved if we throw enough money at them. What we often forget is that not all problems have a solution.

What are the problems of utilizing the world's deserts? Despite the severe heat, human life encounters no serious physiological limits in a desert, provided drinking water is plentiful. Even in the hottest deserts human beings can maintain a normal body temperature by evaporation of sweat, but the water must be replaced, or we rapidly become dehydrated. In a hot desert, the water lost through evaporation may exceed one liter per hour. At such rates a single day in a hot desert without water can be lethal.

There are five stages of human use of the world's deserts. The first, most primitive use is hunting and gathering of food from natural resources, as among the dwindling tribes of South African bushmen and Australian aborigines. Second, plants not directly usable by humans can be fodder for grazing domestic animals, such as camels or goats, herded by nomads. Third, in simple agriculture humans select and cultivate specific plants, either for themselves or for domesticated animals. A fourth stage, modern mechanized agriculture, is achieved when irrigation and machinery are used for cultivation of vast areas. Such developments eventually lead to a fifth stage, gradual urbanization and the growth of big cities.

In my lecture in the main amphitheater of the Sorbonne, I spoke of the interesting physiological adaptations of desert animals and plants, and how these resources can be utilized advantageously by nomads and to a minor extent in primitive agriculture, but I also warned that the need for fuel and the voracious appetite of domestic animals, such as goats, destroy the desert's sparse and slowly recovering vegetation.

I also emphasized that modern agricultural methods, involving ample irrigation, fertilizer, and machinery, can lead to high productivity. Although much land is unsuited for agriculture because it is uneven, rocky, and mountainous, there are large areas that can be used profitably. The problems that arise, I said, are technical rather than biological. Adequate water must be obtained, and energy is needed for pumping and for machinery.

Yet the very irrigation that is essential for high productivity creates problems. Aside from wastage of water caused by poorly designed irrigation systems, irrigation leads to salinization. All natural water contains some minerals, and when the water evaporates, it leaves behind minerals that accumulate in the soil. It may take years, but the accumulated salts will eventually kill all crops, unless the salts can be leached out with an excess of water that can drain away. Chemical treatment of the soil is possible but provides no permanent solution. I mentioned that on the lower Colorado River substantial areas of first-class agricultural land have been abandoned because of accumulated salts. Sections of California are now threatened in the same way.

Moreover, the water supply is limited. In heavily irrigated areas, groundwater is disappearing at fearsome rates. As water is pumped up, the groundwater level steadily sinks, and the cost of pumping increases. Worse, the underground water is only slowly renewed. The aquifers under the Sahara Desert are thousands of years old; the use of this accumulated capital inevitably leads to depletion.

Cloud seeding, in attempts to increase natural rainfall, has proved ineffectual because no amount of seeding can wring water out of a dry atmosphere.

Desalination of sea water is feasible but energy intensive, and practical only in the vicinity of the sea. We again face the familiar problems: the depletion of fossil fuel, the technical and political problems of nuclear energy, and the natural limitations on wind or solar energy.

I spoke with deep regret about the bleak prospects for the future of the world's deserts and cited examples of deterioration, such as the development of resorts and cities in southern California and Arizona that have led to increased abuse of a fragile environment. I described how roads and expanding tourism destroy and litter the countryside. I told how areas otherwise inaccessible can be reached with field vehicles that tear up vegetation and destroy wide areas that may never recuperate once their precarious biological balance has been disturbed.

I concluded my lecture by stating that there is little support for the belief that the world's deserts constitute an unused resource that could be developed to the benefit of humankind.

My audience included many nonscientists, including the French minister of scientific and technical research. He had arrived late, delaying my lecture and keeping the audience waiting. He disagreed with my pessimistic view and, after my lecture, gave a long political speech, expounding on my lack of scientific wisdom. After more than half an hour the audience began to leave; I remained and politely pretended to listen. When his diatribe finally ended, it was a relief to me and to the few friends who remained in the audience. My French hosts seemed embarrassed and clearly shared my view.

In May 1978 Margareta and I visited my daughter Astrid in Minneapolis. When her two children were old enough to be in school, Astrid had continued her education and earned a Ph.D. degree in experimental psychology. She now had an NIH postdoctoral fellowship at the University of Minnesota and had moved there with her family. The year before, Astrid had announced that she was divorcing John and would marry his friend Pete Stewart. I liked John and understood his resentment. However, after some years John also remarried, and both Astrid and he seem happier in these later marriages.

As always we enjoyed visiting Astrid, and I was glad to know Pete better. While we were there, I gave a lecture at the University of Minnesota and spent time with Carl Hopkins, an excellent zoologist who studied the communication systems of electric fish. These fish use electric signals for communication, and each has its own personal frequency for the signals it emits. If two fish that use the same frequency chance to come near each other, they change frequencies to avoid jamming each other's signals.

Seeing the work of other scientists and talking with them is an important part of a scientist's life. One can learn more in half an hour of personal contact than in days spent in a library trying to catch up on new developments. Besides, what gets published is past work, often a year or more old; in discussions of ongoing research one gets new ideas and finds out about current work. Discussions among scientists, whether in a research laboratory or at scientific meetings, are immensely stimulating and absolutely necessary to keep us mentally alert and productive.

The Camel's Nose

IN EARLY JUNE 1978, the
Fourth International Conference on Comparative Physiology took us
to Crans sur Sierre in southern Switzerland. These comparative physiology conferences were organized by Liana Bolis, a professor of physiology at the University of Rome. I had first met her seven years earlier
at the physiology congress in Munich, where she had introduced herself and suggested that we ought to organize an international meeting
to promote comparative physiology.

Within a year Liana had arranged and financed a conference of
forty or fifty scientists from around the world to meet in the small town
of Aquasparta, not far from Rome, to discuss subjects such as the comparative physiology of locomotion and respiration. Every second year
since then she has arranged similar meetings, attended by specialists in
various fields of comparative physiology, selected and invited by an

international organizing committee. Topics have covered a wide range of subjects. Most of the meetings have been at Crans sur Sierre, a mountain resort where Liana owns a chalet and spends brief vacations.

The theme of the 1978 meeting was the comparative physiology of primitive mammals, which include insectivores such as shrews, marsupials such as kangaroos and opossums, and the duck-billed platypus and the spiny anteater.

It is a mistake to assume that the term "primitive" necessarily means simple or inferior. For example, shrews have jaws and teeth similar to those of the earliest mammals, yet they are in no way inferior to their more recently evolved relatives. Because their traits are ancient and conservatively maintained, zoologists prefer to call them conservative rather than primitive traits. And in other respects, such as reproduction, shrews are as advanced as other mammals. A similar range of traits can be seen in the duck-billed platypus (a monotreme mammal), which is highly conservative in its mode of reproduction: it lays eggs! However, in regard to locomotion the platypus, an excellent swimmer, is highly advanced. Nevertheless, for this conference we chose the word "primitive" because it was such an ingrained term and everyone knew what was meant.

From Crans sur Sierre Margareta and I went by train to Bern, where the professor of anatomy Ewald Weibel had organized a small follow-up conference for those who were especially interested in animal locomotion. That stimulating meeting lasted all day, followed by an excellent dinner at Weibel's house. Both Margareta and I were intrigued by a delicious cheese known as Säriswähler and produced only in a local village. When we left on the train for Copenhagen the next day, Ewald presented us with a huge piece of Säriswähler cheese, carefully wrapped in layers of plastic to confine its strong odor.

We seated ourselves comfortably in an empty compartment, hoping that we would remain alone during the long trip. Despite the wrapping, the odor of our cheese was strong, but we rapidly got used to it. Each time the train stopped at a station, more passengers boarded, but whenever anyone opened the door to our compartment, they invariably withdrew again. At a station in Germany a hefty woman pulled the compartment door open, heaved her suitcase onto the overhead baggage rack, and resolutely sat down. She sniffed, looked around, and within seconds jumped up, pulled down her suitcase, tore the door open, and was gone. In Göttingen a Danish woman traveling with her young son came into the compartment. Her first remark, which evidently she didn't realize we understood, expressed strong disapproval of "these dirty old Swiss cars." We didn't reveal the source of the strange atmosphere, and wondered if she would ever again choose to travel on a Swiss train.

We spent two weeks in Sweden with Margareta's parents at their country house in the Stockholm archipelago. Margareta's mother and I sometimes put out fish nets in the evening. She let out the net from the stern of a rowboat that I slowly moved from the beach into deep water. This left the net situated perpendicular to the beach so that it would catch fish that moved along the shoreline at night. We often caught enough fish for both lunch and dinner. Early the next morning we took up the nets and hung them on poles that were set near the edge of the water. Margareta and her father helped remove the fish and clean the nets, which were often badly tangled and torn by struggling fish. We left the nets to dry completely before putting them away. Cotton nets, however well tanned, quickly rot and lose their strength if the slightest moisture remains when they are stored. For Christmas later that year Margareta and I gave her parents a set of nylon nets to replace their old ones. Nylon nets do not rot; besides they are stronger and less visible in the water and therefore catch more fish.

If the weather was good, all four of us went out in a rowboat to fish for strömming, a dwarf race of herring that lives only in the Baltic Sea. We usually fished where we saw other boats or where schools of strömming had attracted flocks of screaming gulls and terns, which dove to catch the fish. We let our boat drift while we threw a line with a jig over the side of the boat, leisurely pulling it up and down. If we were lucky, we could readily catch twenty, thirty, or even more in half an hour.

We also went for long walks in the forest. Margareta's mother is especially good at finding mushrooms where others see nothing. She can identify every species and knows which ones are edible; we always had the delicious reward of a basketful of fine mushrooms for dinner.

That year, although our anniversary fell on a Wednesday, the Midsummer holiday was as usual officially celebrated on the following Friday. Midsummer celebrations are based on traditions that go back to ancient fertility rites. The central element is a tall, slender pole, the trunk of a young birch, with a crosspiece near its top. We covered our pole with garlands of green birch leaves and an abundance of wildflowers. Young women and children tied more wildflowers into two wreaths, which we hung from the ends of the crosspiece. We then raised the pole, and everybody joined hands and danced around it, singing old Swedish folk tunes.

When everybody was exhausted from dancing, we had coffee and large trays of cakes and cookies, served outdoors in the sun. White clouds drifted slowly across the sky as the blue and yellow Swedish flag flew from the flagpole, welcoming the arrival of sunny days and light nights after the long, dark winter.

Margareta's younger sister, Christine, came from Stockholm to join us for the Midsummer weekend. She and Margareta have similar features, but Christine is shorter and her hair redder. She had never been in America, and we arranged for her to travel with us when we returned home.

Christine was independent and readily took care of herself when Margareta and I were at work. She took a bus trip to the coast, where she visited some of my colleagues at the Duke Marine Station. For a Swede, used to swimming from granite cliffs, the endless sandy beaches along the Atlantic coast were a swimmer's paradise.

At this time Margareta was no longer teaching at Duke Preschool. She felt that her educational background in Sweden was inadequate for being a truly excellent teacher; there was so much more to know and learn about, she said. In 1978, after first earning a master's degree at the University of North Carolina at Chapel Hill, she started work there toward a doctorate in education and cultural anthropology. Her course work and research were demanding, but in spite of our frequent travels she earned her degree in 1984.

During this period Margareta also had medical problems. A common complication of diabetes is a high probability of changes in the blood vessels in the retina, which eventually lead to blindness. In 1976 Margareta had had a hemorrhage in her right eye, which within a year led to total loss of vision in that eye.

This development was alarming because of the likelihood that the other eye would become involved. Aided by a new method for blood glucose monitoring that became available for home use in the early 1980s, she started on a meticulously monitored regimen of keeping her blood sugar level as close to normal as feasible. Now every day she takes ten or twelve blood samples and injects five carefully adjusted insulin doses. The threatening changes in the good eye have been reversed, she has no other complications such as renal failure or amputations, and she is able to lead a seemingly normal life.

Another delightful memory from the summer of 1978 stems from a trip we took to Maine to see Mimi and her friend Loyal. The weather was cool and clear, a lovely relief from the heat and humidity in North Carolina. Just outside Ellsworth, halfway up a long hill, we found Mimi's ramshackle house. We turned into a steep, rocky driveway and instantly recognized the old purple-and-yellow panel truck that Loyal had rebuilt as a camper; we had first seen it a few years earlier when he and Mimi came through Durham on their way from California to Maine.

Mimi was thirty, tall and striking, with slow, graceful movements. Her heavy blond hair usually hung straight down to below her hips, but that day she wore it in a single thick braid. She had an engaging wide smile, and people always felt at ease in her presence. Every day Mimi and I went to a nearby bird sanctuary, where we walked barefoot on the soft paths in the quiet forest. We smelled the fragrance of the spruce, we looked at trees and plants around us, and we often thought alike without a need for words. Mimi had a keen sense of other people's thoughts and troubles. She liked Margareta and was clearly happy to see what close friends Margareta and I were.

Bent and his girlfriend, Holly, came up from Boston and stayed for several days. We baked bread and cooked fresh fish for dinner. Mimi had an evening job as a waitress in a fashionable restaurant, so we could spend all day together until she had to leave for work.

One day we went swimming in a deep, abandoned quarry filled with cool, clear water. When I saw Mimi climb high up on the cliffs and dive headfirst into the dark water, I was concerned that she would be in serious trouble if she had an epileptic seizure. Her condition was only partly controlled by medication, and I was always worried about her unpredictable episodes. However, she appeared able to live a nearly normal life.

Another day we went to see a fisherman friend of Mimi's and found him smoking alewives, a fish that looks much like a big herring. We took a huge bagful back to the house and sat in the kitchen, eating the still-warm fish with freshly baked bread.

As we talked about the alewives, Bent said he'd like to catch eels in Eel Pond on Mount Desert Island, a large pond that Bent and Mimi knew well from their childhood summers there. We bought some hardware cloth and wire and spent some hours making traps, which are no more than a cylinder of hardware cloth with an inverted funnel at each end, where the eels can easily get in but have trouble getting out.

We drove to Eel Pond and set the traps, baited with fish offal. Early the next morning Bent, Margareta, and I went back long before the others were awake. Uncertain of the legality of fishing for eels without a license, we went about it quietly. We pulled up the first trap; it was empty and the bait undisturbed. Second trap; empty. Every trap was empty. Why was it called Eel Pond? We didn't try again, and a few days later Margareta and I headed back home.

WE CONTINUED to travel every year as I received invitations to attend conferences, deliver lectures, and spend time with interesting colleagues. My passport was crammed full of stamps. In 1979 alone we traveled to London, Paris, Strasbourg, Budapest, Stockholm, Leningrad, Moscow, and Tel Aviv.

One result of our travels was that I gained some insight into jet lag. Both Margareta and I noticed that we suffer less from jet lag when we travel east, although most people say they suffer less going west. We are both early morning people and are sleepy in the evening. When traveling east, the night is short, but waking early is easy for a morning person. Conversely, an evening person, when traveling east, wants to sleep in the morning and hates to be awakened early. Going west is the opposite: the internal clock of a morning person says that it is well into the night when evening finally arrives, but the evening person doesn't mind remaining active at night.

Of the many places we traveled to, a visit to the Soviet Union remains prominently in my mind. We had been invited to visit the USSR as guests of the Soviet Academy of Sciences; it would be the first trip there for either of us. Visas were required, and tickets could not be bought before the visas were issued. I had expected that as official guests we could readily obtain visas in Washington, but the Soviet bureaucracy there could not oblige us, so we counted on better luck in Stockholm, our departure point for the Soviet Union.

There were long queues at the consulate in Stockholm. We joined the lines, and after hours of waiting we were asked to leave our applications and passports. However, the curmudgeonly official gave us no information about when we could get the visas, or if we would get them at all. Every day I returned and waited in the interminable queues but received no further information until, the day before we were scheduled to leave, our visas were suddenly ready. The binding of Margareta's passport was damaged as if it had been taken apart for photocopying, but we couldn't imagine why her travels for the past several years should have interested the Soviet Union.

Travelers to Leningrad usually went by air, but we planned to take the overnight ferry to Finland and to catch the train from Helsinki to Leningrad. The large vessels that ply between Sweden and Finland carry hundreds of cars and thousands of passengers. The ferries leave in the evening, and the view is beautiful as the ships glide slowly through the narrow passages between islands in the Stockholm archipelago.

In Helsinki only a handful of passengers were waiting for the train to Leningrad. Once under way we discovered that the train had no restaurant or buffet car, but at one end of our car was a tiny cubicle with a hefty middle-aged woman whose only function seemed to be to serve tea. We enjoyed the tall glasses of hot, dark-brown tea she brought us, but we declined the giant-sized sugar cubes she offered us. The woman didn't indicate that she wanted money, but as she spoke only Russian, I didn't find out whether we were special or whether tea is a normal courtesy on Russian trains.

At the Finnish-Soviet border the passport control was simpler than

I expected; the border officials merely glanced at our passports. However, in the next compartment the customs agents spent more than half an hour with two young men. Everything in their baggage was spread out on the seats. Some colorful magazines, including an issue of *Playboy*, were confiscated. Were the publications subversive, or did the officials want them for their own enjoyment? Then two beautiful Finnish knives were discovered and promptly confiscated. I praised our fortune that our possessions hadn't been unpacked; perhaps our visas said that we were harmless or that we were guests and should be treated courteously.

At about eight in the evening our train pulled into the crowded station in Leningrad. I was relieved to see two tall, powerfully built men elbowing their way toward us. The older man presented himself as Professor Vladimir Govyrin and his companion as his collaborator Nicolay Balashov. They took us to the giant Hotel Moskva, and after we were installed on the third floor, Nicolay left us, saying that he would pick us up the next morning at ten.

It was late, and the hotel restaurant was closed, but on each floor a tiny buffet had wine and caviar for sale. I also bought a bottle of mineral water because it was well known, though denied by Soviet officials, that the water supply in Leningrad was contaminated by an intestinal parasitic protozoan, *Giardia*. The mineral water had an unpleasant metallic taste that made me think of rusty iron. Throughout our stay we brushed our teeth with this water, drinking as little of it as possible.

At a desk on each floor of the hotel, a stern woman kept a sharp eye on who came and who left her domain. She knew who we were, accepted our key whenever we left, and without a word handed it back when we returned.

In the morning Nicolay picked us up and took us in an official car to the academy. A well-dressed and personable young man named Ivan was assigned to us as an interpreter. When we got to know him better, he told us that he was a student of Chinese and Chinese history and worked as an interpreter for the sake of the income and other conveniences.

Ivan had been directed to show us the Hermitage, with its priceless art collections, but we said that we would rather see the nearby anthropological museum, which friends in Sweden had told us about. After some discussion he allowed us to miss one of the world's most famous art collections. We walked to the anthropological museum, where Ivan instructed us to follow him closely as he walked briskly past the long queue of visitors waiting patiently to be admitted. Without slowing, he shouted a few words to the guards and passed them while Margareta and I hurried to keep up.

"What did you say to the guards?" I asked.

"I only shouted, 'The delegation is coming,' and didn't slow down for them to ask questions," he replied.

In the museum the old-fashioned displays showed people of Siberian tribes living in tents or earthen huts, with the handcrafted tools and simple household implements they used in their daily activities. These exhibits portrayed a way of life utterly different from anything we are familiar with. We did not regret our choice.

As guests of the academy, we had an official car. Ivan showed us magnificent palaces built in prerevolutionary times for the tsars. Ruined during the German siege of Leningrad during World War II, they had been restored to their original splendor, representing wealth and luxury beyond imagination, with gold and elaborate decorations everywhere. On these visits Ivan always took us past the queues of other visitors.

One day Nicolay Balashov took us to visit the Brain Research Institute, where a small museum featured Pavlov's office as a central attraction. Seeing this great scientist's simple study and imagining his daily life was a moving experience.

At Govyrin's institute, the I. M. Sechenov Institute of Evolutionary Physiology and Biochemistry, there were 250 scientists and 500 other personnel. I gave a lecture that was simultaneously translated. After every few sentences I stopped to let my interpreter take over. I understood enough of what she said to know that she did a fine job; she understood not only my words but also the meaning of what I discussed.

We visited the zoological museum in order to see the director, Dr. Scarlato, who had been at our house a few years earlier while visiting Duke with a group of Soviet scientists. As his name suggested, he was of Italian extraction. During Napoleon's retreat from Russia in 1812, an Italian officer named Scarlato had remained behind, settling in the area and becoming the progenitor of a large family.

Dr. Scarlato showed us a baby mammoth recently uncovered in

Siberia. Much of the skin was still in place, darkened, hard, and brittle like old leather. The black, fibrous material surrounding the bones had little similarity to what once had been muscle. This evidence convinced me that reports about mammoth meat being so well preserved that it remained edible were sheer fabrication.

Galena, one of the scientists in Govyrin's institute, suggested that we go shopping together on her day off, and we spent most of the day with her. The stores had no window displays, but wherever Galena saw people in a queue, she would run up and join them and only then ask what the queue was about and whether it was worth waiting for.

The standard shopping technique required a shopper to join the queue at the sales desk and ask a salesperson for whatever you wanted. If the item was available, you would join another queue at the cashier's window and wait for a chance to pay. Then you had to rejoin the line at the sales desk to pick up the merchandise with the receipt in hand.

Later, in Moscow, we learned how to save time. We were with Nicolay and another scientist when I wanted to buy some Russian tea to take home. We entered an appropriate store, where Nicolay and I joined the queue at the sales desk while our other friend joined the line at the cashier's window, paid, and brought us the receipt; thus we could pick up the tea without another long wait.

In Moscow we visited the well-known A. N. Severtzov Institute of Evolutionary Animal Morphology and Ecology. The person I most wanted to see was Nicolai Kokshaysky, whom I had met at Lighthill's scaling conference in Cambridge in 1975. He is famous for an ingenious method for studying air flow around a flying bird. Tiny helium-filled soap bubbles can be suspended in air, and if the size of the bubbles is carefully adjusted, the weight of the bubbles will exactly match the lift provided by the helium so that the bubbles float in the air, neither sinking nor rising. When a bird flies through air filled with these bubbles, the trail of decaying air vortices left by the bird can be captured on film and later analyzed for an understanding of the aerodynamics of bird flight.

I gave a lecture at the institute and had long and interesting discussions with the scientific staff. I was struck by the contrast between the openness and intellectual curiosity of these scientists and the lack of freedom that was evident in every little detail of their lives.

When it came time to leave Moscow, we went to the airport in an academy car, and I was grateful that Nicolay remained with us. The first hurdle was to check in for our flight. A long queue ahead of us was slowed by the careful checking of all baggage, evidently to prevent smuggling of prohibited goods or valuables.

I had two large tins of the best Russian caviar in my suitcase. I knew it was considered contraband, for it by far exceeded the small quantities that could be purchased for hard currency. Besides, our large tins had been purchased with rubles, a practice that was highly illegal to begin with; and exporting them was even worse.

Nicolay walked directly past the queue to the control point, where I heard him mention something about the academy. One of the controllers immediately grabbed our two suitcases and placed them unopened beyond all the open baggage that was being inspected. We thanked Nicolay for his wonderful help and said goodbye.

At the passport control a stern man behind a glass wall inspected us with a hostile stare. He looked at my passport, examined every page, looked at me with a piercing gaze, looked at the passport again, then at me with the same ugly, hard stare. He kept me standing in front of him for what seemed many minutes, looking alternately at the passport and me. It was an unnerving experience. After I was at last allowed to pass and felt the relief of being free, I realized that the intimidating procedure was designed to unnerve any person who tried to leave the country under false pretenses. Margareta went through the same harrowing experience, and when she was finally permitted to pass she was as shaken as I was. Boarding the Swedish plane, we were happy to leave a country where ordinary people were friendly and kind but the state's officials so hostile.

From Stockholm we flew to Israel to continue the camel studies begun in Kenya two years earlier. With Amiram Shkolnik and Bob Schroter we drove to Kibbutz Qaliah, a settlement north of the Dead Sea in a very dry and hot area several hundred meters below sea level and less than a kilometer from the cave where the Dead Sea Scrolls were found. Working in a kibbutz had advantages. We ate our meals in the common dining room and had all our time available for work, and although we slept on mattresses on the floor, our cabin had air-conditioned rooms.

We knew from our studies in Kenya that camels conserve water by exhaling air at a temperature below the body core temperature. Yet the temperature difference did not seem sufficient to account for the minute amount of water lost from the respiratory tract. Could the exhaled air be less than fully saturated? Could water be removed from the exhaled air on its passage from the lungs and out?

In Kenya we had realized that to answer these questions, we needed to measure the transient changes in the humidity of the camel's exhaled air during a single exhalation. Since then Bob had acquired a tiny humidity sensor, recently developed for use in meteorological balloons. It was only a few millimeters long and ideal for our use because it responded to humidity changes in a fraction of a second. Besides, it was temperature insensitive, a necessary characteristic because the temperature of the airstream changes rapidly during a single breath.

We found that if a camel had not been drinking for several days, the relative humidity of its exhaled air might be less than 50 percent. Clearly, water was removed from the exhaled air as it passed through the upper airways and nose, but what was the mechanism? Later, of course, once we knew the answer, it was so simple that I wondered why I hadn't thought of it long before.

It turns out, as we had suspected in Kenya, that water is extracted from exhaled air only when camels are dehydrated and breathe very dry air. Under such conditions their normal nasal secretions dry up, forming a layer of dried-out mucous and cellular debris on the nasal surfaces. These dried-out secretions are hygroscopic; that is, they take up water vapor from moist air, the way dry crackers or biscuits become moist and limp when the air humidity is high.

When a camel inhales, dry outside air flows over its nasal surfaces, drying them out. During exhalation, as moist air from the lungs flows over these dry surfaces, they absorb some of the water vapor. Thus the air that exits through the camel's nose loses some of its water content and its relative humidity falls below 100 percent.

After returning to Duke I settled all doubts about this simple explanation. My laboratory mechanic and I designed an artificial camel nose scaled to the exact dimensions of the nose of a living camel. We then sent air in and out through this model nose, dry air moving in one direction to represent inhalation of desert air, alternating with moist air flowing in the other direction, representing exhalation from the lungs.

The streams alternated at the same frequency as in the normal respiration of a camel, six breaths per minute.

Inside the model nose was a hygroscopic layer representing the dry nasal membranes. To my immense satisfaction, the surface was dried by the inhaled dry air and took up water from the moist exhaled air. This artificial nose could reduce the humidity of the exhaled air to an amazingly low 5 percent. It was a great relief to explain the old Australian results, and we were pleased to have uncovered an unknown physiological mechanism.

It isn't often that a scientist has the pleasure of discovering something entirely new, and we agreed that our work might merit publication in the *Proceedings of the Royal Society*. Back in my laboratory at Duke, I was enthusiastic about our results and told my collaborators and students about the immense joy of suddenly finding the solution to a seemingly intractable problem that had bothered me for years.

No scientist who makes a discovery wants to be scooped, but Bob and I felt at ease because few scientists were interested in camels, and few or none knew about the peculiarly low evaporation from the camel's nose. Furthermore, it was unlikely that anyone else was aware of the newly invented humidity sensor that was necessary for our work.

Our manuscript was accepted for publication in the *Proceedings of the Royal Society* and would appear in the June 1981 issue. In January that year I got a jolt when reading a favorite comic strip, *Peanuts*, in the morning paper. In the first panel Charlie Brown says to Lucy: "I just found out why camels can go so long without water. It has something to do with their big noses." In the next panel Lucy turns to the dog Snoopy and suggests that with his large nose he could go for years without a drink. We had been scooped! I wrote to Charles Schulz, the artist, to ask how he had learned of our unpublished work. A secretary replied that Mr. Schulz couldn't remember. At any rate, in the scientific literature we retained our priority.

CHAPTER 16

Rewards and Recognition

IN 1978 I received a telegram
from Paris informing me that I had been elected a foreign member of
the French Académie des Sciences. It was a tremendous surprise and
honor to be elected to one of the world's most exclusive scientific bodies, which at that time had no more than twenty-five foreign members.

In the spring of 1979, Margareta went with me to the three-day celebration for new members. I tasted more fine champagne on these
three days than in all the rest of my life. The first day's reception was
held in the academy itself, a luxurious old palace in central Paris across
from the Louvre. The second day's elegant but informal reception was
hosted by the president of France, Valérie Giscard d'Estaing, at the
Palais de l'Elysée. Being greeted personally by the president of France
underscored the remarkable respect in which science and scientists are
held in France and other European countries, where many streets are

293

named after scientists, and not only the most famous. On the last day the mayor of Paris, Jacques Chirac, received us in the Hôtel de Ville. I had not been to a celebration lasting that long since attending a country wedding in Norway when I was eighteen; on that occasion the festivities had still been in progress when I left after three days.

In early June 1980 friends in California called to tell me that Pete Scholander had broken his hip, but they assured me that he was recovering and doing well. Then, on June 13, I received news that Pete had suddenly died; he was seventy-four. I had known him for more than half my life, since 1938, when he came to lecture in Copenhagen while I was a student there. But Pete was more than a good friend; he was my scientific hero. His curiosity and his enthusiasm for nature and science had no bounds. Physiology was his way of life, and his brilliant mind led him to an array of important discoveries.

Pete was a genius at pinpointing important problems and designing novel methods for their solution. He compared the heat production of arctic and tropical mammals. Then he wanted to know how humans keep warm in the cold and studied Eskimos, Lapps, Australian aborigines, and Norwegian students.

Pete wondered why Arctic mosquitoes could freeze to ice in winter and thaw out alive in summer. How could fish swim in polar waters and avoid freezing to ice? He noticed bubbles of air in ice from glaciers and wondered if these hermetically sealed samples could tell us about the earlier composition of the atmosphere. Climatologists would later base their models of past global climate changes on just these bubbles.

In the tropics Pete wondered how the sap could rise to the tops of the tallest trees and how mangroves with their roots in sea water could keep the salts out. He also discovered that hemoglobin speeded up the diffusion of oxygen through water, a phenomenon he called facilitated diffusion. He wondered how gases could get into the swim bladders of deep-sea fish, which live at depths where the pressure reaches several hundred atmospheres.

Among Pete's many discoveries, my favorite is his study of wave-riding dolphins. Dolphins, or porpoises, may playfully ride along just in front of the bow wave of a ship, motionless and seemingly effortlessly remaining in place. This phenomenon had been the subject of theoretical analyses, but Pete thought that the theories, in his words, "missed the boat." On board a Norwegian sealing vessel, he improvised a simple device to measure the forces produced by the bow wave and found that it could indeed provide the forward thrust for a dolphin to remain motionless in the wave. In his article on the phenomenon he characteristically concluded with the words: "This, I believe, is the way dolphins ride the bow wave, and if it is not, they should try."

I was asked to prepare a formal biography for the National Academy of Sciences. I completed this task in six months but did not see it published until 1987, seven years after Pete's death, because the Academy was far behind in the publication of all biographies of its deceased members. While working on the story of Pete's life, I heard from several of his old friends about events that I could not include in the biography. However, these anecdotes tell so much about Pete that I wish to quote them here.

John Pappenheimer, professor of physiology at Harvard, wrote:

In February 1943 or possibly '42 Glenn Millikan, John Lilly, and I went up to Stowe, Vermont, for some skiing. About halfway down the mountain I was surprised to run into Pete Scholander—surprised because I had seen him the month before at Eglin Field, Florida, where I was measuring carbon monoxide emissions from machine guns in fighter bombers and Pete was testing survival equipment for rubber lifeboats.

Pete had been detached from Eglin to help John Talbot test some Army Arctic gear in Vermont, and when I ran or skied into Pete he was absolutely furious because Talbot (a colonel) wouldn't give Pete (only a major) permission to take a couple of enlisted men with him up to the summit of the mountain to test the gear under tough conditions. So I volunteered to go instead.

We camped literally on the very summit of Mt. Mansfield, and the weather turned foul during the night—a three-day blizzard with winds which must have been in the 100 mph range, as they sometimes are in that region in winter. We had food for only one day. I cut my left thumb opening a tin can, and this turned out to be rather important (to me) later on. On the afternoon of the second day we decided to make a descent. It was a bad decision. We soon got lost in the blinding snow and wind and were barely able to put up the tent again (on a steep slope) before our hands got too cold.

The third day broke bright and clear with a temperature of −40° *in the valley,* and God knows where we were. It was during the descent along a wind-blown ridge that both of us froze some parts—including my injured thumb. This became infected after thawing and it was *extremely* painful and looked awful. The local doc filled me full of phenobarb and packed me on a night train to my base in Philadelphia.

One surgeon who saw the thumb was Al Blalock (on a visit from Hopkins) and he thought it should be amputated, but by good luck the first shipment of Penicillium in the form of a green slimy looking solution had just arrived at the Philadelphia General, and they pumped 5 ml of this every hour into my buttocks. This saved the thumb, and I have blessed Alex Fleming [who discovered penicillin] almost every day since, because playing the cello is one of the most important things in my life.

So that is the story of [Pete Scholander and John Pappenheimer] on Mansfield. I wouldn't have missed it for anything, and Susan [Scholander] tells me that Pete felt the same way.

Malcolm Gordon, professor of zoology at the University of California at Los Angeles, described an event that took place in 1959:

Pete, Francis Haxo, Ben Amdur, and I flew up to the Moravian mission station at Hebron, near the entrance to Hebron Fjord, in northern Labrador. There were two primary purposes of the trip: to get information on freezing tolerance and resistance by Arctic fishes living under the winter ice, and to make observations of photosynthetic rates and light compensation levels for macroalgae under the winter ice. We flew by commercial flights from Boston to Montreal to Goose Bay, then by chartered ski plane (I believe the plane was a twin-engine Otter, or something similar) to Hebron.

We had a very successful stay of about ten days at Hebron, then radioed for the plane to come back to pick us up. The plane arrived a day or two later, in very good, clear sunny weather.

After we got all our gear and ourselves on board, the pilot began what initially was a normal takeoff. However, just before we became airborne the righthand ski hit an ice ridge very hard. The impact broke the ski in two, about 6–8 inches back from the point of attachment of a wire cable that stabilized the front of the ski.

Two things then happened simultaneously. The large, heavy broken front end of the wooden ski started flapping around in the slipstream, attached to the wing by its wire cable. It quickly (within a minute or so) smashed several windows next to which we were seated, and also started punching large holes in both the starboard wing and the side of the fuselage. The remaining two-thirds of the ski, still attached to its main pivot point on the landing gear and to another cable to its rear end, was pushed by the slipstream into a vertical position, with its jagged broken end pointing straight down. This produced a tremendous drag on the plane, drastically slowing its air speed and increasing fuel consumption.

The pilot really knew his stuff. He started circling to the right, thus keeping the flapping end of the ski from doing more damage. He also adjusted engine power to compensate for the drag, and told us that he would have to make an emergency landing at a small U.S. base not too far away (a radar station at Saglek Fjord, to the north of Hebron). He radioed Saglek, and we flew there.

On the way Pete was getting us all organized for an emergency landing—however it was to come out. We got all available fire extinguishers, and wrapped ourselves up in sleeping bags and whatever else that was soft and available. We then sat and rooted for the pilot.

The airstrip at Saglek was smooth ice, and the base personnel all came out to stand near the runway, complete with the fire and rescue vehicles they had. Our pilot came down very gradually, touching our tail ski first, then the intact left ski, then, very smoothly, the broken right one. Much to our relief we did not ground loop; the ski simply flattened out when its front end contacted the runway ice, and we coasted to a stop.

We all piled out immediately and spent the next few minutes running around crazily, looking at the damage to the plane, acting like near-hysterical kids, talking like mad and generally feeling relieved. Pete obviously had been *very* worried about the outcome, and repeated a comment over and over that went something like "Yesus Christ, we fooled them again, didn't we? They didn't get us this time." Never before nor afterward did I see him so agitated. He spent the next hour or two in the bar slaking his thirst and talking about how organized religion had been foiled again.

We flew back to Goose Bay later that same day on an Air Force plane.

These events are so characteristic of Pete and his friends that I feel they should be recorded. Pete always seemed to get away with risks that most of us would hesitate being involved in.

Margareta and I went to California for the memorial services at Scripps Institution. Several of Pete's friends spoke; I expressed my admiration for him and his exceptional contributions to science. Ending the ceremony, Martin Kamen, the Nobel-winning biochemist, played several moving violin selections; Pete had often played chamber music with him and other gifted musicians.

LATER IN THE summer of 1980 we were in Budapest for the Twenty-eighth International Congress of Physiological Sciences. At the suggestion of Professor Arisztid Kovach, we had made reservations at Hotel Astoria. Its old-fashioned elegance and impeccable service reminded me of the Britannia Hotel in Trondheim, where I had spent many happy after-school hours with my childhood friend Per.

Before the opening of the congress, the International Union of Physiological Sciences, IUPS, held its General Assembly. The IUPS represents physiologists in about fifty countries, with a membership of more than 25,000. I was one of the five representatives from the United States. On that occasion I was elected president of the IUPS for a three-year term.

In a lecture I gave on the last day of the Congress, I discussed new discoveries about how animals perceive the world around us. The sensory modalities that animals use to obtain information about their environment can be classified in five categories: chemical stimuli (such as taste and smell), mechanical energy (such as hearing), electromagnetic radiation (such as light), electric fields, and magnetic fields. The first three are of course familiar to us, but because the last two are outside the realm of human perception their importance has only recently become clear to us. However, even among the first three there is striking new information that reveals how remarkably different the sensory world of other animals can be from ours.

For many animals, for example, smell and taste have other, previously unimagined roles in communicating and transmitting information. Many animals communicate by releasing pheromones, chemical substances that attract mates, signal alarm, provide orientation clues, and so on. The chemoreceptors that receive these pheromones tend to be astoundingly sensitive. Catfish, for example, are able to detect chemical substances that are important to them in minute concentrations, comparable to dissolving one hundredth of a teaspoon of the amino acid alanine in an Olympic size swimming pool. Salmon migrating from the ocean to the stream where they were born are led by subtle chemical characteristics that distinguish the water of their native stream.

Humans perceive a wide range of sound frequencies, roughly between 50 and 20,000 hertz, or cycles per second, but other animals utilize both higher and lower frequencies. Bats emit sound impulses at frequencies much higher than we can hear and rely on the echoes of these ultrasounds to avoid obstacles and locate insect prey. Similar forms of echo location have been found in some birds and in whales.

Investigators at Cornell University have found that pigeons can perceive frequencies as low as 0.05 hertz, or one cycle per twenty seconds. Because low-frequency sounds are transmitted almost without attenuation over long distances, they are thought to be important in animal orientation and navigation. It is now known, for instance, that long-distance communication between elephants, which the human ear cannot hear, depends on low-frequency sounds.

What we call light is electromagnetic radiation, and our eyes are sensitive to a fairly narrow band of wavelengths. Some other animals are sensitive to shorter and longer wavelengths. Many insects and birds see in the ultraviolet part of the spectrum, invisible to us but important for them. Wavelengths longer than what we call light, infrared radiation, cannot be perceived by the eyes of any animal, simply because each quantum of infrared radiation carries insufficient energy to cause a photochemical reaction. Nevertheless, some snakes, notably the pit vipers, have special organs on each side of the head that can detect infrared radiation from the bodies of warm-blooded prey.

Fish use electric fields. Electric fish emit strong discharges to defend themselves and to stun prey, and weak electric impulses are used for communication. Even fish that do not emit electric signals take advantage of weak electric pulses. A hungry dogfish can discover the minute electric impulses emitted by the breathing muscles of a flounder resting on the ocean bottom, covered by sand. These electric gradients may be as low as 10^{-8} volt per centimeter, equivalent to one pole of an ordinary 1.5 volt flashlight battery placed in New York and the other pole 1,500 kilometers away in Florida.

To humans the most mysterious sensory modality is probably the magnetic sense. Some birds perceive magnetic direction, possibly obtaining information about geographic latitude from the angle between the vector of the magnetic field and the vector of the force of gravity. Perception of the magnetic field may be related to microscopic particles of magnetite, the substance known to ancient mariners as lodestone, that have been found in the heads of pigeons, bees, and other animals.

I enjoyed summarizing all this new information about how animals perceive the world in which they live. Yet what we now understand merely emphasizes how little we know about the complexities of what animals perceive.

WE LEFT BUDAPEST for Denmark and a meeting at a country estate owned by the University of Århus. Two of my old friends, Kjell Johansen and Dick Taylor, had organized the meeting and, with help from the more than fifty comparative physiologists who were there from around the world, made it the occasion for a celebration of my sixty-fifth birthday. Bent, who was then teaching physics and biology at the University of Massachusetts, was there too. The final event was a birthday banquet where outstanding Danish food and much good wine were served. Kjell had invited two members of the Royal Danish Philharmonic to enliven the party. They came dressed in Viking costumes—homespun cloth and horned helmets—and played for us on ancient lurs, long, curved musical horns from the Bronze Age, unearthed from Danish peat bogs.

One very special surprise was a bronze medal that Dick and Kjell had had struck for each participant in the conference. The front showed a stack of my books, *Desert Animals, Animal Physiology,* and *How Animals Work,* surrounded by a camel, a kangaroo rat, a frog, and a desert snail, with a sea gull soaring over them. The reverse bore the inscription:

COMPARATIVE PHYSIOLOGY PERSPECTIVES
CELEBRATING KNUT SCHMIDT-NIELSEN'S 65TH BIRTHDAY
SANDBJERG, DENMARK JULY 22-26 1980

The banquet lasted for hours, with toasting, drinking, and speeches. Margareta kept track of the speeches; when she had counted fourteen, I felt it was time to respond and thank everybody for a wonderful birthday and the many kind things they had said about my contributions to physiology.

BACK AT DUKE life continued as usual. My research laboratory was active, and I was teaching a course in animal physiology that gave me the close contact I liked to have with students. I had declined all opportunities to become chairman of my department; I much preferred not to be involved with administration and had never served on university-wide committees or otherwise participated in administrative activities.

In the late summer of 1981, disturbing news swept the Duke campus. Without the knowledge of the faculty, the administration had decided to build, on land donated by the university, a library and museum for documents and memorabilia from Richard Nixon's presidential years.

The construction of the library would be financed by a private foundation, using funds contributed by friends of Nixon and other private sources. Duke would donate the land, and in return the university would gain access to archival material of great historical interest. The library would be similar to the other presidential libraries. Mr. Nixon had earned his law degree with honors from Duke in 1937, and locating his presidential library at Duke seemed reasonable.

The president of the university, Terry Sanford, announced Duke's plans to members of the Executive Committee of the Academic Council, members of the history department, and other leading faculty members, and by early August the decision to accept the Nixon library and museum seemed nearly final. As more people learned about the plans, controversy mounted. Few faculty members held a favorable view of the former president, and the thought of a museum that inevitably would glorify a man who had been forced to resign from office in disgrace was abhorrent to most.

Mr. Nixon's lawyers indicated that because of simultaneous negotiations with the University of Southern California in Los Angeles, the deadline for the final acceptance was August 11. However, as faculty resistance grew, the alleged deadline slipped by. In early September, when the fall term had begun, the Academic Council voted unanimously to reject any library proposal that included a museum to glorify Richard Nixon, and then voted to form a committee to propose stipulations for accepting a presidential library.

A colleague in the Anatomy Department, Sheila Counce, phoned to say that she wanted to propose me as chairman of the library committee. I protested, but Sheila persevered. An objective report was needed in order to formulate conditions for accepting the archives at Duke, and she was certain I could help produce an unbiased assessment. I also had the necessary academic stature and had never participated in campus politics. Other potential chairmen were perceived as violently opposed to anything related to the former president.

I agreed to chair the committee, whose seven other members did a marvelous job of compiling information. We determined the quantity of archival materials to be housed in the library, which included papers, related textual material, some 30,000 gifts received by the president from foreign and domestic sources, about 500,000 photographs, 4,087 video tapes, and 7,046 audio tapes. There were 950 of the so-called Nixon tapes, which had surreptitiously recorded about 4,000 hours of conversations.

We obtained information from the National Archives about how to house these valuable archives under conditions that would permit their scholarly use. Then we compared our findings with the size of existing presidential libraries and the use of their collections.

Our main charge was to establish conditions for locating the archives at Duke, and we rapidly agreed to recommend two indispensable conditions. One was that control of the archival collections be vested in the university and not in a private foundation funded by Nixon friends and supporters. Second, the library should be limited to archival material and should not include a museum that would glamorize the former president and thus compromise the integrity of the university.

In about five weeks the committee produced and delivered to the Academic Council a well-documented report. The report affirmed the archival value of a collection of presidential papers and evaluated the effect the proposed museum would have on the normal activities of the university. The report concluded that a museum open to the general public would result in conditions incompatible with the orderly functioning of an already cramped and heavily trafficked campus.

These conclusions were based on the experience at the Lyndon Johnson Presidential Library at the University of Texas in Austin. Given the number of visitors to the Johnson Library, which was in many ways comparable to the proposed Nixon library, we could expect perhaps 1,500 visitors per day, and in summer several times as many. There would be a need for new roads and parking areas, hotels, motels, restaurants, and, inevitably, souvenir stands. Once on the campus, many tourists would visit the gardens and the beautiful Duke chapel, which already draws many visitors. The greatly increased traffic to university dining rooms would be incompatible with our obligations to our students.

The committee discussed these potential problems with the university architect, the director of food services, and the chief of campus security forces; all had grave reservations about the effect of greatly increased numbers of visitors to the campus.

Our report was accepted by the Academic Council. It made a clear case in favor of accepting the Nixon archives, but stated that a museum would severely affect the university and compromise its normal functions.

It soon became clear that Mr. Nixon did not want the archival material separated from the contemplated museum, and negotiations with Duke were allowed to come to a stop. The Nixon Presidential Library was finally built in Yorba Linda, California, the former president's birthplace, and opened to the public in 1990.

A similar situation to that at Duke arose some years later at Stanford University in connection with a proposed presidential library and museum to honor President Ronald Reagan. The conditions at the two universities were strikingly similar. Both campuses were already congested and would be severely affected by streams of visitors, Stanford undoubtedly even more so because the San Francisco–San Jose metropolitan area is much larger and densely populated. It appears that the museum question and the inevitable effects on the normal functions of the university again decided the issue—Stanford did not accept a Reagan presidential library. Eventually it was built in a nonuniversity setting, at Thousand Oaks, California.

In 1983, my third year as president of the IUPS, the International Physiological Congress took place in Sydney, Australia. On the opening day I found myself on the stage of the concert hall of the Sydney Opera House, listening to the governor general of Australia addressing the audience of several thousand physiologists from around the world.

Then it was my turn to respond on behalf of the IUPS and its members. On the spot I decided not even to glance at the prepared notes in my pocket. Spontaneous speeches, like scientific lectures, convey more enthusiasm when the speaker, instead of reading from a prepared text, talks to and seeks eye contact with the audience.

That year the Nordic Insulin Foundation had offered financial sponsorship for a lecture series to be named in honor of August Krogh. The Danish Physiological Society had nominated me as the first recipient of this honor, which put me in an awkward position: I was not only a member of the program committee but also president of the union. Although I wanted to withdraw, I realized that my close connection with Krogh and my many years in his laboratory made me the most appropriate person for the lectureship. Furthermore, it might be perceived as a snub if I declined this honor. So I accepted.

I began my lecture with a short biography of August Krogh, one of the greatest physiologists of the twentieth century. Krogh not only made major contributions to fundamental physiology but also devised ingenious new instruments and methods with unheard-of precision. Many of his discoveries are now basic textbook knowledge. He designed equipment to study the effect of carbon dioxide on the oxygen binding of blood and was a collaborator in the discovery of the Bohr effect. He showed that diffusion in the lung is sufficient to explain all gas exchange between air and blood, contrary to what the world's most prominent physiologists then believed. His investigations of gas exchange led to studies of oxygen diffusion from capillaries to the surrounding tissue, work for which he was awarded the Nobel Prize in 1920. He developed methods for measuring the level of human exercise and the rate of blood pumped by the heart, studies that placed his laboratory in the lead in the study of human exercise.

While making fundamental discoveries in human physiology, Krogh maintained an active interest in animal physiology. His doctoral work dealt with the role of the skin of frogs in gas exchange. He studied insect respiration, the swim bladder of fish, and osmoregulation in fish and frogs and insects. He was among the first to use isotopes in biological research, initially in collaboration with George Hevesy. After retirement he studied the physiology of insect flight.

Krogh believed that scientists should derive no personal financial gain from their work. He turned over to his laboratory all income derived from his many inventions, which he never patented because he wanted them freely available to other scientists. Krogh said that he had been greatly helped in his work by what he described as "visual thinking." With the ability to work out in his mind problems of design and experimentation, he could imagine how a novel apparatus would work and make necessary modifications; in contrast, sketches on paper would only impede his freedom to make needed changes.

I found it rewarding to review the many contributions made by this extraordinary man, and it was a privilege to describe discoveries that should be widely known among younger generations of physiologists.

His influence on my scientific development had been great, and after describing his life and work, I finished my lecture with a review of some of my own work.

On the last day of the congress I gave the closing address, listed in the program with the cryptic title "We Live in Medieval Times." In the talk I raised the question of whether the large international congresses remain useful and justify the enormous investment of money and time.

Physiology is an enormous field that tends to splinter into subdisciplines. This trend seems inevitable, and it may even be an advantage because when scientists work in a limited field, there is less that we must know about and keep up with. With the increasing number and size of scientific journals, with more monographs, review volumes, and meeting reports, we are inundated by information.

Complaints about the overwhelming scientific literature are nothing new. More than a century ago, in 1883, the journal *Nature* stated: "Few students of science can fail to feel appalled by the ever-increasing flood of literature devoted to science. . . . New societies, new journals, and independent works start up at every turn."

I added an even older complaint, quoted from the historian DeSolla Price: "One of the diseases of this age is the multiplicity of books; they doth so overcharge the world that it is not able to digest the abundance of idle matter that is every day hatched and brought into the world." These despairing words were written in 1613, long before the appearance of the first scientific journal.

A few years ago a colleague at Duke, David Durack, published an article titled "The Weight of Medical Knowledge." He found that the weight of one year's issues of the *Index Medicus* over a few decades had increased from two kilograms to thirty kilograms. How can we handle this torrent of information? Computers help when we know what to look for, but they cannot find what we do not yet realize we need to know and therefore do not ask for.

I maintain that word of mouth is the most important aspect of scientific communication today, and in this sense we have returned to medieval conditions. When we meet colleagues and friends, it is thoughts and ideas we are eager to hear about. This is the mode of communication that dominated centuries ago; only the speed of transmission today is different. Our scientific forebears traveled slowly by road to the great intellectual centers in Italy and France and remained for years in Pisa or Paris; today we travel by jet to a meeting where the latest discoveries are discussed.

Scientific journals are important for documentation of data, but publication often takes a year or two. Most editors discourage speculation and generalization, even though the thoughts and ideas of talented scientists are immensely stimulating to the rest of us. In addition, specialty journals, with their jargon and acronyms, are increasingly unreadable to the nonspecialist, and important discoveries often go unnoticed by the general reader.

At the large congresses, communication by word of mouth is essential, as in medieval times. Physiologists from around the world gather to hear about new discoveries and to listen as distinguished colleagues give overviews and summarize a lifetime of research experience. In this way they obtain information and knowledge that they could hardly ever glean by poring over the immense number of mostly unreadable journals in which only the specialist can separate important new ideas from masses of trivia. Even if I had all the reprints from the laboratory of a colleague and spent days reading them, it would be difficult to get the essence out of them. In person, a colleague can in five minutes tell me the basic idea of his or her research. Thus, in my closing remarks at the Sydney congress I stressed the importance of bringing people together for personal exchange and communication.

I WAS REELECTED and remained president of the International Union of Physiological Sciences until 1986. My successor, the distinguished British physiologist Andrew Huxley, was a Nobel laureate recognized for his work on the nature of nerve impulses and the mechanism of muscle contraction.

Throughout my presidency I considered it important to support teaching and research in developing countries, and to find means for young scientists from around the world to participate in international meetings and congresses. Both of these goals required financial resources beyond the modest basis on which the IUPS had operated. From outside sources I obtained unrestricted grants that by 1986 amounted to $30,000. In addition, with the help of the IUPS secretary,

Jean Scherrer, and the treasurers, Klaus Thurau and later Heinz Valtin, we obtained $52,500 from UNESCO and the International Council of Scientific Unions to support special educational projects.

I raised the issue of what the IUPS could do to support physiology in less developed countries with Klaus Thurau, who was professor of physiology at the University of Munich. He and I discussed two major problems. One was the tendency toward overspecialization and fragmentation into subdisciplines. For example, neurobiologists were forming separate organizations, endocrinologists were arranging international congresses, and overspecialization was evident everywhere. How could the IUPS counter this threat to physiology as a coherent discipline?

Second, we have too many organizations, symposia, and other meetings in highly industrialized countries, and too few in less developed areas, where research and teaching could be enhanced through regional meetings and workshops. This concern tied into the issue of access to scientific information. With the proliferation of journals and the increasing cost of subscriptions, only the richest institutions in advanced countries could afford to keep up with current developments.

What was needed more than anything else, I felt, was a journal with brief, readable overviews of recent developments in physiology. A journal of this type would allow most physiologists to follow developments outside their fields of specialization and lessen the need to read innumerable specialty journals. Such a publication would counteract the tendency toward splintering, it would help teachers of physiology keep up with new developments, and it would be vital for scientists in areas with inadequate or nearly nonexistent library facilities. Finally, a truly international news journal would keep us all better informed about work in other parts of the world.

Klaus Thurau agreed with these thoughts, and the eventual result was a journal, *News in Physiological Sciences,* or *NIPS* for short, published in collaboration with the American Physiological Society. To make *NIPS* available to physiologists in countries where expensive subscriptions were not feasible, we set the subscription price as low as possible. Once the journal was launched, I succeeded in raising outside funds to subsidize its distribution to institutions unable to pay even the very modest subscription price.

When the Managing Board of *NIPS* asked me to be the editor, I hesitated. The job would be time-consuming; to be a success, *NIPS* had to be attractive, readable, and scientifically impeccable; in other words, it would require a great deal of work. I had other plans: I wanted to write a book I had planned for several years, and a major publisher asked me to head an interesting and financially attractive project. However, I finally accepted because I felt the journal was needed and I wanted to help make it a success. In December 1984 I started inviting and editing articles for the first issue of *NIPS,* to be published in January 1986.

Most scientists who write for journals seem to assume that good scholarship demands formal and complex prose. They forget that good writing is characterized by simple language and common sense. There is no excuse for obtuse writing, for most ideas can be understood if expressed simply, and no law requires readers to work hard to understand what they read. Nearly every article I received for *NIPS* needed editing and revision, and many needed extensive revision. Authors who are told that their articles need clarification seldom know how to improve them, so I had to do most of the editing, at times even rewriting entire articles.

My essential requirement for publication was that I should be able to understand the scientific content of every article we printed; if I couldn't understand it, at least some readers were likely to be in the same position. I also required that all articles be short and concise so that they could be read without great demands on the reader's time. Many of us have put aside an interesting but too-long article "to be read later," which usually means never. Accordingly, the limit was three printed pages (not always strictly enforced).

Many scientists have problems expressing their ideas in a limited space, and everyone seems to believe that masses of supporting data are needed to convince the reader. For similar reasons, most authors also include too many references. The requirements for *NIPS* were that references should cite a few key articles where interested readers could find more information, with a maximum of ten allowed. Many authors sent manuscripts with as many as sixty references, insisting that every one was necessary.

Scientists presumably want others to read about their research, so why do they write long, complicated sentences instead of short, simple ones? Many scientists also favor needless expressions, such as "due to the fact that," which means "because," and "it has been shown that," which usually can be deleted. Consider the sentence "It has been shown that adrenaline accelerates the heart beat, which is due to the fact that it acts on the beta-receptors," in contrast to "Adrenaline accelerates the heart beat because it acts on the beta-receptors." From twenty-two to eleven words in one swoop! With my often drastic editing and rewriting, I expected angry responses, but most authors accepted my revisions without question, and many expressed gratitude for my help in clarifying their message.

My reward for these efforts was satisfied readers. Many commented on the readability and usefulness of the journal. One well-known scientist said to the production manager, Stephen Geiger, that the editor of *NIPS* had an uncanny ability to find only authors who wrote well and clearly.

Having seen so much inept writing, I decided in the spring of 1986 to teach a short course in scientific writing to our research students at Duke. I wanted to show that there is no need for scientific writing to be complex and different from ordinary prose. I have continued teaching this course for more than a decade.

I know that if I lecture to students on how to write clearly, they will listen politely and nod, forget the whole thing, and go about writing as before. Instead I show them how bad scientific writing can be, and how easy it is to prevent people from reading their papers. This procedure suggested the title I gave to my course, "On Falling by the Wayside."

I begin by explaining the importance of the title. To show how easy it is to lose a reader, I ask, "How many titles do you read in a year?" Most students admit they have never given it a moment's thought and are amazed at the answer. Many regularly scan the tables of contents in the widely read weekly journals *Nature* and *Science,* plus several specialty journals. In one year *Nature* and *Science* together contain more than 4,000 titles! Add to that several other journals of interest and a reader might easily scan between 5,000 and 10,000 articles in a year. In that crowd it is easy to overlook an interesting article, unless the title is short and clear.

If you produce a long title with difficult words and acronyms, you are likely to lose a potential reader. You are also apt to lose readers if you use lots of unnecessary words and put them up front, where you should put the most important information. Consider the title "Further Comparative Studies of the Food Competition among Grazing Herbivores." Those first five words say very little and tend to keep the subject of the study less obvious. Also, although "Herbivores" perhaps sounds more scientific, "Animals" is simpler. Why not say "Food Competition among Grazing Animals"?

To show how this actually works, I hand out photocopies of tables of contents from scientific journals and ask the students to scan the titles and see if anything interests them. They invariably stop at short titles such as "Freeze Tolerance in Turtles" or the even simpler "Bloat in Sheep."

It is now clear that, to stop a reader, an author should write a three-line title with lots of difficult words, such as "Partial Coordination of Nervous and Muscular Events Brought About by the Mechanical Effect of Gut Contractions on the Stomatogastric Plexus of *Amphibola crenata*." Moreover, with the increasing use of computer searches, an author can lead potential readers into bypassing an article by using wrong, uninformative, or misleading key words.

I also talk about the importance of the abstract, which represents a second chance to either engage or deter potential readers. Since most readers begin with the first sentence, we must make it interesting. What we are trying to find out should be clearly stated up front. Long lists of data and statistics may look scientific, but they make the whole abstract hard to read. A reader may skip to the last sentence, and this is where the result of the study should be stated in plain words. To demonstrate these points, I hand out photocopies of actual abstracts that seem to raise impenetrable barriers to understanding what the authors accomplished.

During the course I also point out the many other ways of making potential readers fall by the wayside, such as huge tables, complicated graphs, and heavy, turgid prose in the main body of the paper.

In my last session I show slides, donated by colleagues who have actually presented them in scientific lectures. One favorite is a huge table about which the lecturer remarks, "I know this can't be read in the back of the room, but—" and then proceeds to read all the numbers aloud. I show examples of other useless visual "aids"—cracked slides, crowded slides, faded slides, illegible slides, and handwritten slides.

Students claim that the course has helped them to improve their writing. Once they are aware of common shortcomings in much scientific writing, they rapidly discover how readily they fall into the same habits. My greatest reward, however, came one day when a colleague told me that one of my former students had criticized his writing and explained how easily it could be improved.

CHAPTER 17

A Hat, a Ring, a Cannon

For years Mimi's health had me increasingly worried. She never complained, but her epilepsy was not well controlled by medication, and it was evident that her condition was worsening.

In the early 1980s Mimi moved from Maine to Boston, where she worked at various jobs and was fiercely determined to take care of herself. During this time she took the anatomy course at Harvard Medical School and had thoughts of going to medical school. She applied to Duke's medical school but was not accepted.

Mimi was worried that the medication prescribed for her epilepsy to replace the Dilantin she had taken for many years was affecting her immune system. She found several small lumps in her breast that she ascribed to the medication. She underwent biopsies and a partial mastectomy; as she had feared, the lumps were malignant.

313

The week after the initial surgery, she was scheduled for a bone scan before a radical mastectomy was scheduled. I called her almost every evening. She was staying at home in her apartment but was unwilling to say much about how she really felt. She said that she had problems with fluid soaking through the bandages, and I was certain she was in pain even though she would not admit it.

I spoke with Mimi on Wednesday evening, August 22, 1984, the day before her bone scan. When I called again on Thursday evening she didn't answer her phone. I was concerned and wondered if she was back in the hospital. On Friday evening there was still no answer, so I called Bent and asked if he knew anything about Mimi.

Bent, who had been of great help to Mimi during the years they both lived in Boston, went to her apartment. He found the door locked, her car outside, her mailbox full of mail, and her cat hungry and crying, so he called the police.

Mimi was found dead in her bedroom, where she apparently had fallen during an epileptic seizure, hitting her head hard against the wall. Bent phoned me, and Margareta and I immediately flew to Boston. I talked with the medical examiner, who said that the fall might have caused a dislocation in the neck, but he considered an autopsy unnecessary. I took care of other formalities and arranged for the cremation of Mimi's body.

Mimi's ashes were later buried in Maine, a locality she loved more than any other place she had lived. Her closest family and friends gathered there and set Mimi's urn among evergreen trees in an unmarked grave.

Losing a child is an unspeakable sorrow. Even more than a decade later, it is difficult for me to accommodate my feelings of loss. Words cannot convey what a lovable and gentle person Mimi was and how much I wish she were still living.

ONE DAY IN December 1984 I was sitting at my desk, reading my mail. I opened a letter with Swedish stamps, which informed me that the Faculty of Medicine of the University of Lund had voted to confer on me an honorary doctor of medicine degree and wanted me to attend the award ceremonies on May 31, 1985.

I stared at the letter, stunned. At first I didn't understand the emotional impact. Why should an unexpected honor affect me that strongly? Then I remembered; forty-nine years earlier, on the same date, my father had been awarded the same honorary degree at Lund. Tears came to my eyes as I remembered how pleased Father had been. On his way home from the ceremonies he had stopped in Oslo, and over dinner at a small French restaurant he told me about the magnificent ceremonies surrounding the formal award of the degree. How delighted he would have been had he lived to know that I received the same honor half a century later.

Lund is a small university town in southern Sweden that I knew well. I had often visited Father's sister, Dagny, who lived there with her husband, Eilert Ekwall, professor of English at the university. In 1865, some 120 years before my award, my maternal grandfather, Vilhelm Patrick Sturzenbecker, had received his doctorate from Lund, also on the same date, the last day of May.

Through correspondence with the university marshal came inquiries pertaining to the formal insignia of the degree, a doctor's hat and a gold ring. "We must trouble you to let us know your hat size in centimeters, and exact head measurements (which can be taken by your hatter)." Well, I had no hatter, but with a little help I got the circumference of my head as well as the width and length of my skull, which I assumed was what a hatter would measure. I had Father's doctoral ring from Lund, of heavy gold and decorated with an Aesculapian snake surrounded by oak and laurel leaves. It fitted my left ring finger perfectly, so I used the calipers in my laboratory machine shop to measure it to the nearest thousandth of an inch, and for the benefit of the Swedes I recalculated it to millimeters.

For the ceremony I was to wear full evening dress with tailcoat and white tie, but with a black vest in place of the otherwise obligatory white vest. However, for the evening banquet I was to wear the regular white vest.

The award ceremonies took place in the magnificent old cathedral, which dates from the year 1145. With its rounded vaults and romanesque arches it clearly antedates the later medieval gothic cathedrals. The day before the ceremonies, all degree recipients met in the cathedral to be instructed about which chair to occupy, where to step, and how to respond. Every detail was to be precisely correct.

On the day of the formal event the weather was ideal, warm and sunny. Faculty members and recipients of doctorates assembled in University House and were lined up in the proper order for the academic procession. Then, clocked to arrive exactly on time, the procession moved with measured steps the few hundred meters to the cathedral. On both sides were crowds of townspeople. Off to one side on a sward of grass was a small detachment of artillery personnel with several field cannons. When we filed in, the cathedral was already filled with spectators.

The order in which the degrees were awarded was determined by ancient rules. First came the faculty of theology, the oldest university faculty, with the faculty of law second. The faculty of medicine was third, and the faculties of philosophy and humanities came last.

The entire ceremony took place in Latin, and the excellent tutoring the day before helped me understand much of what was said. I wasn't surprised that the theologian spoke Latin, but I was impressed by the eloquence with which everyone else spoke this ancient academic language.

That year the faculty awarded three honorary medical degrees. When my turn came, I was called by the promotor, who latinized my first name as Canutus but left my family name unchanged, as it was too awkward for Latin. I rose and walked the few steps to the platform, where I stood facing the promotor, Stefan Mellander, who spoke in fluent Latin. When he came to the words *annulum aureum*, I knew that he was referring to the gold ring he held between two fingers. I raised my left hand, and he placed the ring on my finger, where it stopped halfway down. He continued his Latin incantation in a slightly less secure tone as he pushed harder, but the ring didn't move.

The promotor gave up, and the ring remained in the middle of my finger while he proceeded to the most impressive part of the ceremony, the placement of the doctoral hat on my head. The traditional old hat is like a cylindrical top hat, but it has fluted black silk on the sides and an Aesculapian snake in a golden oval of oak leaves at the front.

The promotor lifted the hat with both hands and stood on tiptoe. I was too nervous to notice whether his Latin speech continued as he slowly lowered the hat onto my head. Fortunately, the hat fitted, and the moment it touched my head, a cannon shot reverberated from outside the cathedral. I was curious how the system operated with such perfect timing, but nobody seemed willing to divulge the secret.

After the ceremony, I put Father's ring on my finger and wore it until my own ring could be modified. Evidently I had erred in recalculating inches into millimeters, confirming my ingrained dislike of nondecimal-based measuring systems. The goldsmith who remade my ring did a superb job and every detail remained intact. My name and the date, 31 May 1985, are engraved in the ring; otherwise it looks exactly like Father's.

MORE SORROW greeted us later that summer. In July Margareta and I received a letter explaining that Niels Gunder Knudsen had died suddenly, at sixty-eight. Niels had been stricken by a massive heart attack as he carried a large ceramic sculpture of a fox to be fired in a colleague's kiln.

I still miss Niels and his enthusiasm for his art and the living world around him. When I look at the many sculptures we have in our house, I think back to the happy summer visits when Niels and I talked for hours about animals and art.

WHEN I WAS a child, birthdays were important. After I grew up I never paid much attention to them, but in 1985 my colleague Steve Wainwright said he wanted to celebrate my seventieth birthday. He wouldn't reveal much about his plans; he only implied that he would make it an event never to be forgotten.

Steve is more than a zoologist on the Duke faculty; he is a sculptor and an adjunct professor in the School of Design at North Carolina State University. He has a gift for the unconventional and takes great pleasure in doing what others do not even think of. He invited our colleagues from the Zoology Department and their spouses, many students, and other friends—altogether about a hundred people—to a party. Astrid and her husband, Pete, drove down from Washington to join in the celebration.

On my birthday, when Margareta and I drove up to Steve's large house, we saw a rhinoceros grazing on the vast lawn. Steve had borrowed the life-sized fiberglass beast from a friend in Chapel Hill.

Lying on the grass farther down the slope were two immensely long balloonlike structures in colorful plastic that were kept inflated by air pumps. The larger, a 200-foot-long cylinder nearly 8 feet in diameter, was gently curved like a snake. The other structure was an immense sphere with a room-sized annex. Each had an opening at ground level where people could creep in through a sleeve that kept air from escaping and the balloons from collapsing. Inside there was ample room for dozens of people to stand, and the soft, curved walls in psychedelic colors gave an eerie feeling of being conveyed to a different world. The impression was reinforced by the music of a student bluegrass band inside the structure.

Steve's next-door neighbor was Paul Gross, the former Duke vice president and chemistry professor who had given me my initial appointment at the university. Steve, Margareta, and I walked over to visit Dr. Gross, who had celebrated his ninetieth birthday only a few days before. He was working in his garage and greeted us warmly. He remembered me well. "You were an impatient youngster at that time," he said, harking back to my first days at Duke. It was true: I wasn't always tolerant of unnecessary bureaucracy; but what little we had to deal with at that time was nothing compared with the regulations and paperwork that eat up the time of young investigators today. After half an hour we returned to the party, hungry and ready to eat.

Two whole roasted pigs were awaiting us, prepared by local specialists in typical North Carolina barbecue. In addition, two long tables held an enormous selection of meat, smoked fish, salads, cheeses, homemade bread, foreign delicacies, cakes, fruit, ice cream, and more desserts.

After all this wonderful food came the greatest birthday surprise. Steve's daughter, Jenny, and four of her classmates performed a modern ballet barefoot on the fresh grass, a graceful presentation that the young women seemed to enjoy as much as their delighted audience. Jenny, who was graduating from Duke that year, told me afterward that her choreography was inspired by one of my books. How could I ever wish for a more magnificent birthday gift?

The birthday coincided with my official retirement from Duke; now I no longer had any teaching duties. I had always enjoyed my teaching, although invariably I was nervous before a lecture, and more so when talking to students than when lecturing to an audience of scientists. The one advantage was that I always was well prepared for my lectures.

That spring I had offered my last class in animal physiology. Because this was her last chance, Margareta audited the course. I had one condition for accepting auditors: I required them to take all exams. I didn't mind if they did poorly or failed, but they had to try. Margareta was no exception; she took all exams, was graded like the other students, and passed the course.

Margareta's familiarity with physiology has made her an invaluable editor, as when she helped with the fourth edition of my animal physiology textbook. She knows enough physiology to understand and evaluate the text, but not so much that she understands the poorly written or inadequately explained material. She not only is conscientious but also has an excellent feel for good and clear writing. When the book was ready to go to press, Margareta designed the cover, which showed four white dolphins jumping in graceful curves on a deep blue background.

In 1985 I was honored by election to the Royal Society of London and was formally admitted in a dignified afternoon ceremony. At the evening banquet that followed I learned that the society dates back to 1645 and that its official name is The Royal Society of London for Improving Natural Knowledge.

The president, who presided at the admission ceremony, was the physiologist and Nobel laureate Andrew Huxley, whom I knew well from his service on the council of the IUPS during my last term as president. When my name was called and I rose, he took my hand and with a brief speech declared me admitted to the society. To complete my admission, I then signed the old Charter Book of the Royal Society, using a special pen-and-ink set placed next to the open book. To minimize the chances for accidental blotches from the old-fashioned pen, another foreign member, Glenn Seaborg, and I had been subjected to a special practice session earlier that day.

The Charter Book is a folio volume of fine vellum, bound in red velvet with gilt clasps and corners. It has been in continuous use since 1662. No fellow is admitted until the book is signed, unless the council grants an exemption for sufficient cause. Exemptions from signing can be granted if a fellow is physically incapacitated or "if the Fellow so requesting is not resident at a place within a radius of 100 miles of the

Society's apartments." Three centuries ago, a 100-mile trip was a formidable undertaking, requiring a great deal more time and effort than did the several thousand miles I had traveled for the privilege of signing.

The first page of the Charter Book carries the signature of King Charles II. Samuel Pepys recorded in his diaries that he was present when the king signed his name and under his signature wrote the word "Founder." A complete facsimile edition of the Charter Book has been published with all the famous signatures reproduced, some more legible than others. Among the first are Newton, Fahrenheit, and Celsius. Farther on is Charles Darwin, who signed the book in 1839. Among the names from this century are some I knew from my student days in Denmark: Niels Bohr, August Krogh, and Kaj Linderström-Lang.

O VER A LONG career I have received many meaningful honors and awards. A few stand out above the rest, however. One was my election to the Royal Society of London. Another special occasion took place in the fall of 1992. That September we were in Denmark, drinking afternoon tea with Karen Krogh, widow of August Krogh's son Erik, at her summer house. Her telephone rang, and when she picked it up the caller spoke English. She thought it was the wrong number and was about to hang up when she heard the word "Duke" and realized that it was a transatlantic call. My department at Duke had received a fax from Japan announcing that I had been awarded the International Prize for Biology. Would my wife and I be able to attend the award ceremony at the Academy of Sciences in Tokyo at the end of November?

What an incredible honor. The prize was first established in 1985 in honor of Emperor Showa (in the West commonly known as Hirohito), who was himself a distinguished biologist. I would be the eighth recipient of this award, joining an extraordinary group of scientists from around the world.

From the moment we arrived in Tokyo in late November, we were treated as honored guests. An official, Mr. Masanori Hasegawa, met us at the airport and took us to the New Otani Hotel, where we stayed in a luxurious suite. With its 1,700 rooms and 17 restaurants, shopping arcades, and a small American-style grocery market in the basement, the hotel was like a small town.

The day before the formal ceremony, Mr. Hasegawa gave us a three-page protocol covering every minute of the event. Attached were floor plans for the auditorium and the stage where the ceremony was to take place. Mr. Hasegawa politely suggested that we memorize everything carefully.

On the morning of the award ceremonies, at precisely 9:20, Mr. Hasegawa came to the hotel to take us to the Academy of Sciences. Outside, photographers were waiting, and while the cameras flashed, we were escorted into an elegant waiting room. A horde of photographers came into the room and clicked their cameras as a steady stream of scientists, diplomats, ministers, and officials came to greet us.

We knew to the precise minute every event in the program. The first would be a practice session for the ceremony. While we were waiting, a gift from Emperor Akihito was brought in, a giant silver vase of elegantly simple design. On one side the mirrorlike surface was broken by the emperor's insignia, a stylized chrysanthemum. More photographers entered the room to take pictures of the vase and of me holding it.

Margareta and I each had a giant yellow chrysanthemum pinned on our clothing; then we were ushered off for my twenty-minute practice session. We were escorted to the stage of the large auditorium and shown our seats. Several dignitaries and government representatives were already there, and each person went through every move in the correct order and timing for the entire ceremony. At this time all instructions and practice speeches were in Japanese, and at the end of it all I wasn't certain I could remember the exact moments to rise and bow to the emperor, move and give my acceptance speech, return to my seat, and bow again to the emperor.

The next point on our schedule, at 11:11, was our introduction to the ministers and other dignitaries. Then, at 11:18, came our presentation to the emperor and the empress in a special reception room. The imperial family is highly respected in Japan, and Empress Michiko is immensely popular.

The empress, slightly built and dressed in a beautiful silk kimono, was a gentle person with great charm. She spoke fluent English, was easy to talk with, and was well informed about Margareta's Swedish background and interest in teaching. She said only a few words to me but spoke at some length with Margareta. The emperor also spoke excellent English and asked me about my work, but so softly that at times he was difficult to understand. However, I felt comfortable with him, for he was kind and friendly and utterly unassuming.

We were then escorted to the auditorium, where we waited for the ceremonies to proceed. Our introduction to the emperor had taken less time than scheduled, so we were seated alone on the platform for several minutes before the ministers and other dignitaries, at 11:29, entered and found their proper places. At 11:30 the imperial couple entered while the audience respectfully remained standing until their majesties were seated.

I came through the program with no major mishaps, rising and bowing to the emperor and empress at the correct times, walking to the table at the front of the platform to accept the gifts of scrolls and medals, bowing to the emperor on my return to my seat, and listening with my earphones to the translation of the speeches by the emperor and the minister of education. At the right moment I rose to deliver my precisely timed acceptance speech, bowing to the emperor when rising and again when returning to my seat. I noticed that several of the dignitaries listened to me with their earphones, so my speech was being translated.

At exactly 12:05 the master of ceremonies announced: "Their Majesties will leave the hall." While everybody was standing and bowing, we were escorted to our initial waiting room. Ten minutes later Mr. Hasegawa led us to the lower level, where the audience from the ceremony was assembled. I again faced the forest of cameras while people came to congratulate and talk with me.

When the emperor and empress entered the room the cameras did their work again, and I presume that my conversation with the emperor was on TV, for a microphone had been attached to my lapel.

One of the last people to come up to me was a young man who introduced himself as the emperor's chamberlain. He asked about our schedule, for the empress wanted to invite us to the Imperial Palace for tea. When would be a suitable time?

Mr. Hasegawa escorted us back to our hotel, but the celebrations weren't over. That night we had an elaborate dinner with the Committee for the International Prize for Biology. The next two days were occupied with an international symposium on comparative physiology. On the last evening Professor Keiichi Takahashi, the organizer of the symposium, invited a few special friends to his home for a banquet of Japanese delicacies.

During this time Mr. Hasegawa was busy discussing our invitation to tea at the Imperial Palace. We then learned that the emperor's chamberlain had advised Mr. Hasegawa that the emperor, himself a biologist and a specialist in Japanese fish, was so interested that he would rearrange his official schedule and invite us for dinner at the Imperial Palace. This was not a simple move, for it involved not only the imperial household but also negotiations through the Ministry of Education.

It turned out that the only day the emperor could rearrange his official schedule was our last day in Japan. What more splendid conclusion could we wish for to end this fabulous trip?

At precisely 6:20 our limousine picked us up at the hotel to take us to the palace, which is located in the Imperial Gardens, a huge area in the middle of Tokyo. We were expected, and a barrier was silently slid open after we were identified. It was dark, and we couldn't see much, but I liked the sound of the wheels on gravel as we drove toward the palace.

At the door we were received by three men who led us to a room where we could leave our coats. We had asked if it would be appropriate to bring some small token gift, and were told that the emperor would appreciate a photograph of me. Margareta brought three intricately folded paper stars she had made for the empress. We knew that gifts are never handed directly to the emperor; they are always given to an attendant. It was a pleasant surprise to be told that we could hand our gifts directly to the emperor and empress.

After a few minutes an adjutant appeared and led us through a music room with a grand piano and a harp and into a sparsely decorated living room. Emperor Akihito appeared, accompanied by Empress Michiko and their twenty-three-year-old daughter, Princess Sayako, the youngest of their three children. They greeted us with a handshake in the Western manner and asked us to sit.

After about ten minutes the double doors to the dining room were opened, and we sat down to an informal dinner. I sat next to the empress and across from the emperor, who asked about my work with camels and how sea birds manage with no fresh water to drink. Princess Sayako sat at the end of the table, but it was easy to carry on a conversation with her about her studies of kingfishers in the Imperial Gardens. She was studying their nesting behavior, using fiber optics to observe them in the deep burrows in the river banks where they nest.

Sitting in the living room after dinner, I asked the princess if she knew the limerick about kingfishers. She did not, so I recited it:

> Kingfisher, kingfisher blue,
> I should like to go fishing with you,
> but my mother would scream,
> if I fell in the stream,
> so I think I better not do.

The princess and the empress laughed heartily, so I asked the princess if I should write it down for her. I added that I knew another limerick about birds, a well-known one about pelicans, and she wanted to hear that also. Although I knew it well, suddenly I couldn't recall the second line, so I said I would send it to her later.

The empress told Margareta that all family members write poetry, and each month the emperor gives them a theme to write about. The empress has translated Japanese poems into English and gave Margareta one of her books of poems about animals.

When we took our leave that evening, I felt that we could continue talking with these soft-spoken and intelligent people for hours, discussing science, customs, writing, and the arts. Our three hosts walked us to the anteroom and waved goodbye from the front steps. The moment we left the palace, the missing line of the limerick came back:

> A funny bird is the pelican,
> his beak can hold more than his belly can,
> he can store in his beak,
> food for a week,
> I'll be damned if I know how in hell he can.

After we returned home, I sent the princess the complete limerick. Because she wasn't familiar with this popular verse form, I told her about its origin and early history. Some time later I received a long, handwritten letter from her, thanking me for the limericks and for the interesting evening we spent together.

My OFFICIAL retirement from Duke, now more than ten years ago, has brought little change in my schedule of office work and traveling for lectures and meetings. Before retiring I had been gradually reducing my research activities, feeling a strong obligation to free research funds for younger scientists, but I have continued my informal contacts with students and still teach about the pitfalls of scientific writing. I am often invited to lecture at scientific meetings and at other universities, and when it is the students who invite me it is difficult to decline.

At times I have been asked about the future of physiology and what important discoveries I foresee. I have no idea what the future may bring, except that we will have more answers and ever more questions to answer. During the fifty or sixty years since I was a student, I have lived through more "future" in physiology than most of my readers, and I can safely say that nobody would have predicted much of what has taken place. Who could have anticipated Hodgkin and Huxley's work with the squid axon, revealing the nature of the nerve impulse, which was totally mysterious until then? Or who would have anticipated that we would study single ion channels and their molecular configuration? When I was a student we didn't know about channels, and not even a genius can predict the clarification of phenomena whose very existence is unknown.

As scientists begin their inquiries, asking the simple question "why," we seldom know what practical applications our findings might offer. In my lifetime the range of discoveries and their applications has been extraordinary.

I remember sitting on my mother's lap when the family doctor came into the room with his little black bag. It must have been in 1919, when I was four years old and a worldwide flu epidemic killed millions. I was suffering from pneumonia, and I clearly remember the doctor coming to listen to my lungs with his stethoscope. He knew that he could do nothing more, except wait and hope for the best. A physician back then could diagnose a disease and explain its likely course but could offer little in the way of a cure. At most he could offer some encouraging words to make Mother feel better. His little black bag contained no sulfa drugs or penicillin or anything else of much use. Luckily, by 1919 medicine had advanced beyond bloodletting and purgatives and other interventions that either scared patients into getting well or hastened their demise.

Since that time medicine has changed at a rapidly increasing rate. In the 1920s insulin was discovered to treat diabetes, and vitamin deficiency was linked to rickets; in the 1930s came sulfa drugs; and in the 1940s penicillin and other antibiotics were developed. Today kidney and liver transplants and open heart surgery are commonplace, procedures not even dreamt of when I was a student.

Has the study of animal physiology contributed in any way to such advances? Perhaps not directly, but I like to think that many medical researchers gain some insight and some understanding from animal physiology that may broaden their perspective on fundamental medical problems.

In academic and government circles the debate continues over the value of basic versus applied research in a time of scarce resources. I was fortunate to live through a period when it was relatively easy to obtain support for basic research. Projects that were reasonably well conceived and planned were likely to receive funding. Today the competition for funds has become extremely keen; a substantial part of a scientist's time is spent writing grant applications, only to have them unfunded though recommended for revision and resubmission, with more effort spent on honing every detail finer and finer.

The thought that goes into providing detailed research plans and schedules is not all wasted. Knowing that a plan will be scrutinized and criticized by their peers undoubtedly helps scientists crystallize their ideas and clarify their prose, but if every step is circumscribed and outlined in detail, where is the freedom to explore new ideas as they come up? If our scientists could spend more time on science and less on writing applications, their ideas might be more innovative and their lives more creative.

I am also mindful of the extraordinary privilege represented in these pages—of the luxury of travel and my exceptional good fortune in receiving a lifetime investigatorship from the National Institutes of Health. Without the freedom to pursue new ideas and without the help of my many gifted research students and postdoctoral associates, I could never have achieved much.

Today I can in good conscience steer toward a research career only those young persons who possess an irresistible curiosity, an over-whelming urge to devote all their efforts to solving problems, and a high degree of tolerance for frustration and discouragement.

My own success in the field of animal physiology has basically come from a combination of luck, the privileges of my upbringing, and my own simple-mindedness. I am not made for difficult and complex problems. My interest in animals came to me naturally, and the questions I have tried to answer in my research have all seemed simple. Looking back at it all, I inevitably think of the fortune teller in India, who not only seemed to know about my past and the importance of the letter *M*, but also correctly predicted a long life and much happiness.

Acknowledgments

In WRITING this book, I have
come to realize the tremendous gratitude I owe my parents. They gave
me the benefit of a privileged childhood and permitted me to find my
own bungling way to an advanced education. In short, their support
was the foundation for what in the end enabled me to spend a rewarding life in science.

Before I started writing, I compiled a list of hard facts based on old
calendars, diary notes, passports, correspondence, and, especially, long
letters to my parents, which my sister saved and later returned to me.
This gave me the skeleton to build on, although many important events
in my life are left out, including some of which I am not particularly
proud.

My special thanks go to all the many students and colleagues who
over the years have been my collaborators and friends. A few I have had
the courage to ask to read and comment on my drafts. I owe a special
debt of gratitude to two close friends, Frank Dutra and the late Stan
Fletcher, who patiently read my very first drafts; to my colleague in the
English Department at Duke University, Reynolds Price; and, espe-

cially, to Keith Brodie, who at a critical point encouraged me to revise my draft and find an editor, supported by a grant from the Devonwood Foundation.

Writing and revising the book has been a long and exacting exercise, and I am deeply grateful for all the generous help I have received. The original one-thousand-page draft was reduced to seven hundred pages with the help of A. J. Mayhew, and reduced further, to five hundred, by Georgann Eubanks. After I found a publisher, I profited greatly from helpful suggestions by my editor, Laurie Burnham of Island Press, and by my copy editor, Ann Hawthorne. I am especially grateful to Dan Sayre, editor in chief at Island Press, for much help and advice guiding the final stages of editing and book production. In the end, of course, I myself am fully responsible for the final content and form of the book.

Index